Installation Mapping Enables Many Missions

The Benefits of and Barriers to Sharing Geospatial Data Assets

Beth E. Lachman, Peter Schirmer, David R. Frelinger,

Victoria A. Greenfield, Michael S. Tseng, Tiffany Nichols

Prepared for the Office of the Secretary of Defense
Approved for public release; distribution unlimited

CD-ROM of document included

NATIONAL DEFENSE RESEARCH INSTITUTE

The research described in this report was prepared for the Office of the Secretary of Defense (OSD). The research was conducted in the RAND National Defense Research Institute, a federally funded research and development center sponsored by the OSD, the Joint Staff, the Unified Combatant Commands, the Department of the Navy, the Marine Corps, the defense agencies, and the defense Intelligence Community under Contract DASW01-01-C-0004.

Library of Congress Cataloging-in-Publication Data

Installation mapping enables many missions : the benefits of and barriers to sharing geospatial data assets / Beth E. Lachman ... [et al.].
   p. cm.
  Includes bibliographical references.
  ISBN 978-0-8330-4034-3 (pbk. : alk. paper)
   1. Geographic information systems—Government policy—United States.
2. Geospatial data. 3. United States. Dept. of Defense—Information services.
4. Information commons—United States. 5. Military geography. I. Lachman, Beth E., date.

G70.215.U6I57 2007
355.6'880285—dc22

2007018561

The RAND Corporation is a nonprofit research organization providing objective analysis and effective solutions that address the challenges facing the public and private sectors around the world. RAND's publications do not necessarily reflect the opinions of its research clients and sponsors.

RAND® is a registered trademark.

Published 2007 by the RAND Corporation
1776 Main Street, P.O. Box 2138, Santa Monica, CA 90407-2138
1200 South Hayes Street, Arlington, VA 22202-5050
4570 Fifth Avenue, Suite 600, Pittsburgh, PA 15213-2665
RAND URL: http://www.rand.org/
To order RAND documents or to obtain additional information, contact
Distribution Services: Telephone: (310) 451-7002;
Fax: (310) 451-6915; Email: order@rand.org

Preface

Installations and environment (I&E) geospatial data assets are being developed, used, and shared for many different Department of Defense (DoD) missions, including installation management, homeland defense, emergency response, environmental management, military health, and warfighting. There are many benefits in effectiveness and efficiency to using and sharing such data. However, there are also barriers that limit the widespread use and sharing of such assets within and outside DoD, including security concerns, lack of on-going high-level program support, lack of data-sharing policies, and lack of any rigorous analysis to prove the benefits of sharing. This monograph assesses the mission effects of sharing I&E geospatial data assets within the business domain and across the business, warfighting, and intelligence mission areas of the DoD Global Information Grid (GIG). It also analyzes the barriers to sharing and recommends some ways to overcome them.

This monograph should interest those wishing to use and share geospatial data for DoD missions. It should also interest government policymakers and managers who would like to learn more about geospatial data sharing and use across their respective enterprises. A CD containing the full document in color is enclosed at the end of this monograph.

This research was sponsored by the Office of the Secretary of Defense and was conducted within the Acquisition and Technology Policy Center (ATPC) of the RAND National Defense Research Institute, a federally funded research and development center sponsored by the Office of the Secretary of Defense, the Joint Staff, the Unified Combatant Commands, the Department of the Navy, the Marine Corps, the defense agencies, and the defense Intelligence Community.

For more information on RAND's ATPC, contact the Director, Philip Antón, by email at ATPC-Director@rand.org; by phone at 310-393-0411, extension 7798; or by mail at RAND Corporation, 1776 Main Street, Santa Monica, California 90407-2138. More information about RAND is available at www.rand.org

Contents

Figures

Tables

Summary

From the Office of the Secretary of Defense (OSD) to workcenters on military installations, there is widespread use of geospatial information contained in digital databases, specialized software applications, documents, web services, and even hard copy maps for diverse functions and missions. Installations and environment (I&E) geospatial data assets not only support mission areas in DoD's business domain—including emergency response, environmental management, and facility and infrastructure planning—they also support the warfighting and intelligence mission areas.

The widespread use and sharing of I&E geospatial data assets yield many benefits, such as cost and performance efficiencies. Moreover, they can help decisionmakers manage other assets better, enable faster responses for time-sensitive decisions, and improve the communication process across diverse agencies. If data are shared, different organizations can save time and money by not having to develop and maintain the same data; they also avoid problems relating to inconsistencies and quality differences in the data. Using out-of-date or poor-quality data can affect the outcome of a decision or a mission using those data. Many of these effects are very real yet are difficult to quantify or measure.

To encourage the use and sharing of geospatial data assets, DoD and the Office of Management and Budget (OMB) have issued guidance and directives that stress the need for coordinating, sharing, and integrating geospatial data assets across DoD and other federal agencies. In July 2004, within the Deputy Under Secretary of Defense Installation and Environment Business Transformation (DUSD/I&E (Business Transformation)) directorate, a new organization—the Defense Installation Spatial Data Infrastructure (DISDI) Office was created to help facilitate the sharing and integration of I&E assets.

The objective of this RAND study is to assess the net effects of sharing I&E geospatial data assets within the business domain and across the business, warfighting, and intelligence mission areas of DoD's Global Information Grid (GIG) and to recommend how the DISDI Office could maximize the benefits of such sharing. For the study, RAND researchers interviewed over 100 producers and consumers of geospatial data assets, reviewed geospatial and effect assessment literature, and examined sample geospatial data assets to identify the range of missions and effects to them from current and potential future use of these assets. They also developed a methodology for assessing the mission effects of sharing such assets, using it to estimate some effects across DoD. In addition, barriers to sharing were identified and recommendations were made for how DISDI could help overcome such barriers.

What Is Shared, Who Is Sharing It, Why, and How

One of the most common and fundamental geospatial data assets is GIS (geographic information system) datasets. GIS is a class of software for managing, storing, manipulating, analyzing, visualizing, and using digital geospatial data. Geospatial data assets also include other products using geospatial data, such as software applications, documents, hard copy maps, and videos. Geospatial data software applications range from general GIS-based tool sets to simple and sophisticated mission-specific web-based applications.

U.S. military installations across the world are developing, using, and sharing I&E geospatial data assets. Most of the Services' basic digital geospatial data are created, updated, and maintained at the installation or regional level. Historically, the mission functional staff members who needed the data created, maintained, and updated them; for example, Department of Public Works (DPW) staff develop data on building and road infrastructures. Many installations develop and maintain hundreds of GIS data layers, with datasets at different levels of scale and time periods, often maintained because of different needs.

Because of advances in enterprise software capabilities and the growing realization of the benefits of sharing data, installations and the Services are taking a more centralized approach to developing and maintaining basic geospatial data assets. Some fundamental data layers, such as base boundaries, roads, buildings, imagery, and training range areas, are widely used and needed for gaining broad situational awareness across an installation. Therefore, each Service has identified (or is in the process of identifying) basic data layers to be used and shared by organizations across an installation in what is known as a Common Installation Picture (CIP). The idea is to have one map or set of geospatial data shared across each installation.

Service headquarters, functional commands, and regions also develop, maintain, and update geospatial data assets. Other DoD staff, such as DISDI and the National Geospatial-Intelligence Agency (NGA), also exploit I&E geospatial data assets; as an example, DISDI has created the DISDI Portal, a web site where DoD users can view and learn about Service I&E geospatial data. DISDI and other OSD organizations currently focus more on software applications and rely on the installations to supply them with the basic I&E geospatial datasets for those applications. Such organizations may also generate some strategic geospatial datasets, especially ones designed for looking across a region, a nation, or the world, such as a georeferenced point dataset showing installation locations in the world. The NGA develops geospatial data assets for the warfighting and intelligence communities.

But the Services are more than repositories or even managers of geospatial data assets. Each Service has headquarters geospatial organizations to facilitate the development, sharing, and use of geospatial data assets. They facilitate sharing within their respective Services by setting Service I&E geospatial data policies, by being a Service point of contact for geospatial data requests (which they usually forward to the appropriate Service organization), and sponsoring the development of Service-wide geospatial data web viewers so that many military users can access I&E geospatial data assets. Each office also participates with the DISDI Office efforts to establish a DoD-wide I&E geospatial community.

These Service organizations essentially are developing a spatial data infrastructure (SDI) for each of their respective Services. An SDI encompasses policies, standards, and procedures for organizations to cooperatively produce and share geographic data. Components of an SDI usually include institutional arrangements, policies and standards, data networks, technology, users, data, databases, and metadata.

DISDI serves a similar function for the business functions within DoD. It focuses on the business processes, people, and policies necessary to provide installation visualization and mapping capabilities. DISDI is not in the business of creating information technology (IT) systems; rather, it fosters mechanisms by which geospatial data stewarded by DoD installations can be shared with validated stakeholders to help meet their critical installation visualization and geospatial requirements.

DISDI's first major initiative was developing the Installation Visualization Tool (IVT) for the 2005 Base Realignment and Closure (BRAC) process.[1] The IVT was designed for "situational awareness" in the BRAC process and provided a way to view imagery and geospatial data in a consistent fashion for 354 sites, including training ranges, meeting BRAC 2005 threshold criteria. IVT data were used to support other analyses as part of the BRAC process, but no analysis was performed using the IVT data alone.

Through the efforts of DISDI, the Service headquarters, Major Commands, and the installations themselves, geospatial data assets are widely shared among many organizations. For example, these data assets are shared among regional and headquarters levels within the Services. Geospatial data assets are also shared across different Services and other DoD organizations for such mission functions as joint facil-

---

[1] Technically the IVT program office started the development of IVT in 2003, then IVT was transformed into a task of the DISDI Office in July 2004.

ity and environmental management, joint training, warfighting, and intelligence.

We found that there is also a large amount of current and potential sharing with other federal agencies outside DoD and with state and local governments that need geospatial information to assist with key governmental functions such as homeland security, environmental management, and disaster preparedness. Further, because of industry outsourcing, public-private partnerships, and other arrangements, I&E geospatial data are also shared with commercial entities to conduct infrastructure management. Finally, we also found that DoD organizations have a need to share with universities, nongovernmental organizations, and allied governments. Sharing, of course, is a two-way street and DoD organizations need to acquire other government agency data and industry data, such as that from utility companies.

I&E Geospatial Data Assets Enable DoD Business Functions and Warfighting and Intelligence Mission Areas

We identified 12 mission areas, based on traditional installation operations, for which I&E geospatial data assets are now being shared or have the potential to be shared:

- base planning, management, and operations
- emergency planning, response, and recovery
- environmental management
- homeland defense, homeland security, and critical infrastructure protection
- military health
- morale, recreation, and welfare: enhancing quality of life
- production of installation maps
- public affairs/outreach
- safety and security
- strategic basing

- training and education
- transportation.

To illustrate how the I&E geospatial data assets enable different business missions within different parts of DoD, we present diverse examples in the text and an even richer set of examples in the appendix. In this summary, we offer three abbreviated examples of mission support for different organizational levels using the asset.

- **Installation level use:** At the installation level, I&E geospatial data assets have been used to help develop, assess, manage, and operate numerous installation assets, from buildings to natural resources to training ranges. For example, at Fort Hood, Texas, the range GIS aerial and topographic data are used in tank and aviation simulators, which help orient the soldier and saves valuable time on the training range. For A-64 Apache helicopter training, it has cut the amount of time that pilots need to spend on the gunnery range by about one-third.
- **Office of the Secretary of Defense application:** Various offices within OSD use I&E geospatial data assets to help in their strategic analysis, planning, management, and operations. Many of these applications are more recent and are taking advantage of IVT data. The OSD Health Affairs TRICARE Management Activity (TMA)/Health Programs Analysis and Evaluation Directorate has been developing a GIS-based "Military Health System Atlas," to help examine and assess military medical capabilities and their populations. This OSD office uses I&E geospatial data assets in this atlas to help with decisions about medical resource allocation.
- **Uses by other parts of DoD and organizations outside DoD:** Other parts of DoD, such as NGA, and organizations outside DoD, such as state and local governments, also use I&E geospatial data assets, especially for emergency response and homeland defense/security missions. With U.S. Geological Survey support, NGA has the federal lead on Project Homeland, a collaborative effort to provide geospatial information to federal, state, and local

government agencies for homeland planning, mitigation, and response so that the U.S. government can more effectively respond to incidents—whether a terrorist attack or a natural disaster.

I&E geospatial data assets also support different warfighting missions across DoD, including:

- command, control, communications, and computer (C4) systems
- logistics
- warfighting operations
- strategic planning, policy, and assessments.

Here we provide only one example for the warfighting mission that is related to deployed operations but many more are provided in the main document. I&E geospatial data models and techniques are used at forward bases and sites to help build, manage, and operate these sites, such as helping address force protection, critical infrastructure, and other safety concerns. Sharing geospatial expertise help saves money and time and improves safety and planning to help save lives. For example, the Assessment System for Hazardous Surveys (ASHS) program, a GIS-based application software tool to assess capacities for explosive safety hazard reduction, has been used to help plan and manage explosives safety at deployed host nation bases supporting operations in Afghanistan and Iraq.

Assessing Mission Effects

Not only do I&E geospatial data assets aid in a wide range of mission areas, they also generate many different types of mission effects. As we will show, these effects are seen at all levels within DoD—from an individual office on an installation to the Office of the Secretary of Defense. Our definition of effects is broad and includes the attainment of desired outcomes by the individual organization developing, using, or sharing the assets and any other outcomes to any organization from that asset development, use, and sharing. We identified four categories of effects from using and sharing geospatial data assets:

- changes in efficiency
- changes in effectiveness
- process changes
- other mission effects.

At least implicitly, we are suggesting that the goal of use and sharing is to improve the efficiency and effectiveness of organizations' efforts to attain mission objectives, although in some instances, it may have an even more direct and immediate bearing on mission attainments. Organizations often invest in geospatial data assets with the expectation of efficiency effects, in the form of time savings or cost avoidance, and effectiveness effects, in the form of new or improved outputs and outcomes, such as improved operations and decisionmaking. Table S.1 provides some examples of effectiveness effects for different mission areas.

However, organizations often underestimate the extent of those gains. For example, once the data and related systems are in place, organizations often identify additional uses that improve efficiency even more, or they find that the intended use of the geospatial data assets generates benefits that were never anticipated, such as improved communications between two offices or automating a formally manual process.

We were asked to help the DISDI Office identify a methodology for assessing the net mission effects of developing, using, and sharing geospatial data assets across the GIG. We recommend applying a methodology that consists of three elements:

1. Construct an information flow model to understand the range of organizations using and sharing an I&E geospatial data asset.
2. Apply a set of logic models to map how the inputs, activities, and outputs of an organization's data development, use, and sharing lead to outcomes for different customers.

Table S.1
Sample Effectiveness Effects from Using I&E Geospatial Data Assets, by Mission Area

Mission Area	Sample Effectiveness Effects
Base planning, management, and operations	Better placement and siting of new facilities, such as buildings Improved infrastructure and facility construction, management, and oversight Better use of construction and maintenance resources
Emergency planning and response and homeland defense and security [a]	Improved planning and response decisionmaking by having more accurate and common situational awareness of potential and actual incidents Faster response times Better coordinated response with other federal, state, and local agencies Better pre-placement and use of resources
Environmental management	Improved environmental quality, such as reducing erosion and improving water quality Protecting habitat, species, and cultural resources while maintaining installation operational flexibility Reduction in noise complaints More effective working collaborations with community and other stakeholders to address environmental issues
Training	Improved siting of a training range or testing facility by minimizing safety and environmental effects Increased operational flexibility at a training range Increasing the number of hours that a training range or testing facility can be used Cutting by one-third the time on a training range Being able to use more of the installation for training
Warfighting operations	Improved management and operations of base camps and other forward operating sites (FOSs) Improved force protection and safety at base camps and other FOSs More rapid deployment and better use of resources in deployments Faster and more accurate assessments of adversary operations, such as insurgency attacks in Iraq Improving postconflict reconstruction by providing tools for infrastructure reconstruction and management

[a] Since these mission areas have some of the same effectiveness effects, they were grouped together here. See details in the appendix for each mission area's application.

3. To the extent possible, when the data are available, employ a variety of methods for quantifying the logic models.

The information flow model diagrams organizations and the geospatial data assets that they share to understand the institutional structure. This is the first step to understanding how geospatial data assets are shared. Along the way, each organization may see one or more effect. The second step is to apply logic models, which illustrate how the inputs and activities of an organization potentially lead to beneficial outcomes—in other words, logic models illustrate the underlying logic of an organization's activities in relation to an intended end state.

Figure S.1 presents a logic model for some of the geospatial activities of the Camp Butler Environmental Management Program in Okinawa. This logic model shows that the program uses geospatial data

Figure S.1
Logic Model for Camp Butler Environmental Management Program

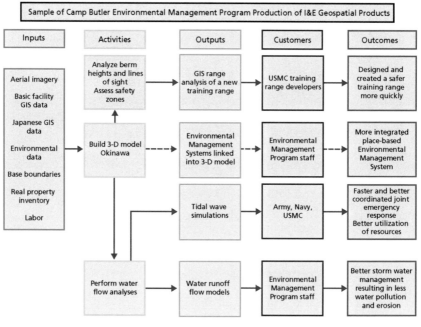

that are collected and managed on the installation as well as data provided by the local community. One key intermediate activity is the creation and maintenance of a 3-D model of Okinawa, which is used to help support watershed modeling, training range development, and tsunami simulations. The activities and outputs of this program support different customers and mission areas (under the outcomes), including training, emergency response, and environmental management. For example, the tsunami simulations improved joint planning and emergency response training, which results in better joint decisionmaking, coordination, and communication; faster response times; and better use of resources for an emergency incident; whereas the water runoff flow models and analysis helped the environmental staff better manage storm water runoff so that there is less pollution and fewer erosion problems from the runoff, such as by more efficiently placing the technologies to capture and treat oil runoff from parking lots. The dashed line indicates that one output based on the 3-D model is planned but has not yet been implemented.

The third step is to apply methods for quantifying some of the effects using the logic models when data are available. We were able to collect some data directly from personnel we interviewed at different installations. In other cases, more complete data were available because other researchers have conducted a cost-benefit analysis or done related research.

In our interviews with Langley AFB personnel, we found that two construction-related functions—dig permitting and delivery order processing—saved from 450 to 2,600 man-hours annually by using geospatial data assets in place of more labor-intensive manual activities.

Aberdeen Proving Ground and Naval Air Station Patuxent River conducted benefit-cost analyses that provide a more complete picture of the net effect of using and sharing geospatial data assets. In 1992, the DPW at Aberdeen Proving Ground conducted a cost-benefit study for the implementation of a GIS for different mission uses across the installation and estimated a net present value of $3 billion in 1992 dollars over an eight-year period. Most of the benefits were in the form of monetized workload reductions. The Patuxent River study took a much broader view of its IT investments and assets and valued its net benefit at \$82.5 million in 2000 dollars. Those mission functions involving geospatial data assets accounted for about \$1.4 million in annual gross benefits.

Using these quantitative measures and DoD's *2005 Base Structure Report*,[2] we show how a relatively straightforward extrapolation can produce a ballpark order-of-magnitude estimate of the total potential effect across all DoD installations in the United States.[3] We present the results of extrapolations in Table S.2 and provide more detail on the calculations in Chapter Seven. Note that these results are very rough approximations and rely on a very small sample set and the original estimations provided by persons we interviewed and studies reviewed. Thus, we suggest that although these are estimations, they help to convey a sense of how large the potential annual benefits may be, measured narrowly in terms of mostly workload reductions.

Table S.2
Order-of-Magnitude Estimation of Potential Annual Benefits of I&E Geospatial Data Asset Use at DoD U.S. Installations

	Annual Savings at the Installation	Extrapolation to All DoD U.S. Installations
Langley dig permitting and construction order processing	450 to 2,600 hours per year	About 100,000 to 600,000 hours per year, which is equivalent to 50 to 300 full-time personnel[a]
Aberdeen Proving Ground gross benefits	\$1.3 million (2005 dollars)	\$200 million (2005 dollars)
NAS Patuxent River gross benefits	\$1.7 million (2005 dollars)	\$360 million (2005 dollars)

[a] Assuming a 2,000 hour work year, the savings equate to between about 50 and 300 full-time personnel.

---

[2]   Department of Defense (2005a).

[3]   We base the extrapolations on the assumption that total plant replacement value (PRV) and total base personnel (both of which are reported in the *2005 Base Structure Report*) are roughly proportional to the amount of savings realized at an installation. If an installation such as Patuxent River accounts for about 0.5 percent of all U.S. PRV and personnel, we assume that its benefits account for about 0.5 percent of all potential U.S. savings from using geospatial data assets. In this estimate, we used only U.S. installations, since operations at installations in other parts of the world may have different procedures.

Benefit-cost, return on investment (ROI), cost-effectiveness, and cost-avoidance analyses can be powerful decisionmaking tools that provide quantitative measures of certain types of effects—mainly efficiency gains, such as time and dollar savings. But because they are computationally complex, time-intensive, and not easily updated, such methods are probably not feasible to use by themselves on an ongoing basis to measure and monitor the full effects of efforts of the DISDI Office and other organizations to promote the use and sharing of geospatial data assets. Our methodology of using together the information flow models, logic models, and such quantifying methods (when feasible) provides a more appropriate tool for assessing effects.

Likely Future Use and Sharing of I&E Geospatial Data Assets

The development, use, and sharing of I&E geospatial data assets continue to grow for many reasons. The data and technology are now easier to use in more user-friendly ways, such as in web-based systems; standards and interoperability conditions are being implemented that help facilitate use and sharing by multiple organizations and individuals; efficiency and effectiveness benefits are being realized, which helps facilitate investment in these resources; sharing is mandated by OMB Circular A-16; and centralized military organizations, such as the Service headquarters offices and DISDI are now facilitating the use and sharing of such assets.

Because of these factors, the use and sharing of I&E geospatial data assets across the GIG will likely continue to increase. We identified some likely future trends in several mission applications. First, there will likely be more use by the warfighting and intelligence communities. The relationship between these communities and the installations will evolve because of the benefits in collaborating to improve the speed and effectiveness of the U.S. military's ability to rapidly deploy and respond where needed around the world to fight the Global War on Terrorism as well as perform other missions, such as providing humanitarian assistance.

The second trend is increased demand and use of I&E geospatial data assets by other parts of OSD and DoD. The demand is driven by the benefits to decisionmaking and management that result from integrating, aggregating, and sharing geospatial information from installations to higher management, in such areas as real property, environmental issues, military health capabilities, and safety. Sharing will also likely increase with NGA because of the need to coordinate all types of geospatial information across all of DoD and the growing use of I&E geospatial data assets to support warfighting and intelligence missions.

A third trend is the increased demand for nonmilitary community geospatial data by DoD agencies and for I&E geospatial data assets by nonmilitary communities. Military installations want and need access to local, state, and federal data to help perform their missions. For some organizations, such as the U.S. Army and Air National Guard, such sharing with state and local governments is critical to their mission. Likewise, other U.S. government agencies need geospatial information to help with key U.S. government functions, such as homeland security, environmental management, disaster preparedness and response, and land-use planning. And at the local level, military installations share their I&E geospatial data assets with adjacent local governments to help with joint infrastructure, utility, safety, and natural resource management and for emergency planning and response.

Finally, a fourth trend is the evolution of geospatial applications toward web-based spatial applications, using more real-time information, and integrating and sharing more detailed information from diverse sources.

Despite these trends, we have identified a number of barriers that continue to impede successful sharing of I&E geospatial data assets. The main ones identified in our study are

- security concerns and other data restrictions
- different IT system, firewalls, and policies
- lack of communication or collaboration between different functional organizations and disciplines

- lack of knowledge about, interest in, or expertise to use I&E geo-spatial data assets
- lack of data-sharing policy, standards, and contractual agreements
- reluctance of data stewards to share assets, fearing that they will lose control over access to their data
- lack of on-going high-level program support and investments
- risks from sharing undocumented, poor-quality, and out-of-date data.

Such barriers will need to be addressed to realize significant increases in the future use and sharing of I&E geospatial data assets across the GIG. DISDI and the Service geospatial information offices are playing an important role in addressing such barriers.

Recommendations

In April 2006, NGA was formally identified to OMB as the lead office for DoD geospatial information management issues. We offer a number of recommendations for how DISDI, in partnership with NGA, can do even more to help DoD overcome the barriers to I&E geospatial data asset development, use, and sharing. The first set of recommendations relate to policy. The DISDI Office serves an important role in setting OSD policy regarding I&E geospatial data assets. DISDI should collaborate with NGA to provide more official OSD policy guidance about the need to share geospatial data assets, about security concerns, and about how to share assets, such as by providing guidance about developing memoranda of understanding/agreement (MOUs/MOAs) for data sharing.

The DISDI Office also has an important role in coordination and outreach regarding I&E geospatial data asset development and sharing within as well as outside DoD. The DISDI Office has already done a lot to help coordinate and conduct outreach across DoD about the need to share and how to share. DISDI should continue and expand on coordination and outreach efforts inside DoD, assist OSD organizations in their acquisition and use of I&E geospatial data assets, cultivate a close working partnership relationship with NGA, and expand outreach and coordination outside the DoD.

Since standards, contracting, and quality control processes are all key to the sharing of I&E geospatial data assets, DISDI has an important facilitator role in such processes. First, it should help develop and promote I&E geospatial data standards development and adoption. It is also important that DISDI provide OSD policy and standard contracting language for military contracts that involve digital geospatial data and analysis.

The tasks mentioned above represent quite a large workload for the current DISDI staff. DISDI presently has a director and four contracted staff members and some limited funds to allocate for projects. Given such tight resources, it is important that DISDI is managed wisely. We present three recommendations. The first is to examine the benefits and feasibility of temporarily expanding the DISDI staff, perhaps using knowledgeable geospatial staff Intergovernmental Personnel Act (IPA) assignments from other DoD organizations part time at DISDI.

Second, to help assess its success in promoting data sharing, DISDI should apply the methodology we developed for assessing effects, i.e., using together information flow models, logical models, and, when feasible, cost-benefit analysis and other quantifying methods. DISDI can use this approach to help understand, assess, and explain the full range of effects from the development, use, and sharing of I&E geospatial data assets. Such assessments can be used to help DISDI manage its current and future investments.

Last, we recommend that DISDI establish processes for managing future investments by applying the Government Accountability Office (GAO)[4] maturity model. Long-term improvements in processes, policies, and organizational relationships can be planned systematically using the IT Investment Management (ITIM) maturity model developed by the GAO.

---

[4] Effective July 7, 2004, this agency's name changed from General Accounting Office to Government Accountability Office.

Conclusions

U.S. military I&E geospatial data assets are being shared with many different organizations in many different ways inside and outside the DoD. The assets support many mission areas—from the installation level to the Office of the Secretary of Defense. The effects from such use and sharing relate to both efficiency, such as cost and manpower savings, and effectiveness, such as improving operations and decision-making. There are also secondary benefits, such as improving communications and working relationships. However, the use of I&E geospatial data assets in many of these areas has just begun and more needs to be done to fully accrue such benefits across the GIG. Data asset use and sharing, and the benefits, will likely increase and reach even more users within DoD. However, barriers exist to such sharing. The DISDI Office and the Service geospatial information offices serve an important role in addressing the barriers to data asset sharing to facilitate more I&E geospatial asset development and sharing across the GIG.

By implementing the methodology suggested here to help show the benefits of geospatial data sharing and the policy recommendations outlined for the DISDI Office, I&E geospatial data asset development and sharing will continue to increase and to accrue significant financial and operational benefits across the GIG helping to improve mission performance and ultimately save lives and dollars.

Acknowledgments

We would like to thank the Defense Installation Spatial Data Infrastructure (DISDI) Office for sponsoring this report, particularly Colonel Brian Cullis, USAF, former Executive Manager of the DISDI Office.

Our study also gained important insights from discussions with DISDI staff, Service GIOs, and I&E geospatial asset developers and users throughout and outside DoD. We would like to thank the many individuals who supplied us with information, including Mark Alexander, Will Allen, Paul Allred, Daniel Andrew, Kevin Backe, Rich Bannick, Rob Barber-Delach, Glen Barrett, Bruce Beard, Marc Beckel, Nicholas Beltramelli, Chris Bendall, Jay Berry, Richard Bilden, Mary Brenke, Andrew Bruzewicz, Stephen Bryce, Rusty Bufford, Gene Burchette, Mike Burks, Barton Clements, Leann Cotton, David Cray, Vicky Cwiertnie, Josh Delmonico, Kelly Dilks, Patrick Easton, Mark Eaton, Craig Erickson, Kelly Ervin, John Esposito, Dan Feinberg, Julie Finnegan, Lou Garcia, Jane Goldberg, Bill Goran, Lisa Greenfeld, Jeree Grimes, Tom Haake, Mark Hamilton, Andy Hanes, Keith Harless, Eric Harmon, Jo Hewitt, Vance Hoyt, Jim Huisenga, Antwane Johnson, Karen Jones, Steven Kestler, Susan Kil, George Korte, Amii Kress, Greg Kuester, David LaBranche, Dat Lam, Andrew Lambert, Dave Lashlee, Bob Lepianka, Steve Luttrel, Andy Marotta, Mike McAndrew, Patti McSherry, Linda Moeder, Bill Mullen, Robin O'Connell, Rich Olivieri, Robert Opsut, Fred Pease, Matthew Pittman, Francis Railey, Roy Rathbun, Terry Rhea, Mark Riccio, Charlene Rice, Ed Riegelmann, John Robilliard, Andrew Rogers, Bill Russell, Jim Sample,

Heidi Santiago, Mary Pat Santoro, Christopher Scott, Tobi Sellekaerts, Kenneth Shaffer, Laura Silsbee, Dan Silvernale, Brandi Simpson, Marilyn Slater, Bradley Smith, Deke Smith, Denise Smith, Kieren Smith, Chris Stanton, Bill Stevens, Tom Terry, Mike Thomas, Hal Tinsley, Greg Turner, Brian VanBockern, Dan Vernon, Joe Vogel, Scott Walker, James Wassenberg, Roger Welborn, Dan Wheele, Nathaniel Whelan, David Wiker, and Gary Wolfe.

The final monograph has benefited greatly from reviews and comments by several knowledgeable people, including John Moeller and David Oaks. In addition, numerous RAND colleagues made substantive, editorial, graphical, and administrative contributions to this monograph.

Any errors of fact or judgment that remain are solely those of the authors.

Abbreviations

AAAV	Advanced Amphibious Assault Vehicle
AB	Air Base
ACC	Air Combat Command
ACHP	Advisory Council on Historic Preservation
ACSIM	Assistant Chief of Staff for Installation Management
ACUB	Army Compatible Use Buffer
ACZ	airfield clear zone
ADUSD	Assistant Deputy Under Secretary of Defense
AFB	Air Force Base
AFEUR	Air Forces Europe
AF/ILEPB	Civil Engineer of the Air Force, Programs Division, Bases and Unit Branch
AFSPC	Air Force Space Command
AKO	Army Knowledge Online
AMC	Air Mobility Command
ANG	Air National Guard
ANSI	American National Standards Institute
AOR	area of responsibility
APG	Aberdeen Proving Ground
APZ	accident potential zone

ARNG	Army National Guard
ASHS	Assessment System for Hazardous Surveys
AT	anti-terrorism
AT&L	Acquisition, Logistics, and Technology
ATPC	Acquisition and Technology Policy Center
AVL	automated vehicle location
BASH	Bird Aviation Safety Hazard
BCEG	Base Closure Executive Group
BIM	Building Information Model
BNOISE	Blast Noise Prediction computer program
BRAC	Base Realignment and Closure
C4	command, control, communications, and computers
C4I	command, control, communications, computers, and intelligence
CAA	Clean Air Act
CAC	Common Access Card
CADD	computer-aided design and drafting
CALFEX	Combined Arms Live Fire Exercise
CAPP	Contingency Aircraft Parking Planner
CATS	Consequence Assessment Tool Set
CDF	California Department of Forestry
CDTR	Chemical Demilitarization and Threat Reduction
CENTAF	Central Command Air Forces
CENTCOM	Central Command
CERCLA	Comprehensive Environmental Response, Compensation and Liability Act
CERL	Construction Engineering Research Laboratory

CHaMP	Community Health and Medical Program
CIO	Chief Information Officer
CIP	Common Installation Picture/Critical Infrastructure Protection
CMC	Commandant of the Marine Corps
CMTC	Combat Maneuver Training Center
CNI	Commander Navy Installations
COBRA	cost of base realign actions
CoI	community of interest
CONUS	Continental United States
COP	common operational picture
CTA	Central Training Area
DASD	Deputy Assistant Secretary of Defense
DASH	Deer Aviation Safety Hazard
DASN(IS&A)	Deputy Assistant Secretary of the Navy for Infrastructure Strategy and Analysis
DCIP	Defense Critical Infrastructure Program
DDESB	DoD Explosives Safety Board
DEM	digital elevation model
DGINet	Distributed Geospatial Intelligence Network
DHS	Department of Homeland Security
DISDI	Defense Installation Spatial Data Infrastructure
DISR	DoD Information Technology Standards Registry
DLA	Defense Logistics Agency
DoD	Department of Defense
DoD-GEIS	DoD-Global Emerging Infections System
DOE	Department of Energy
DOI	Department of the Interior

DPW	Department of Public Works/Directorate of Public Works
DTRA	Defense Threat Reduction Agency
DUSD	Deputy Under Secretary of Defense
ECM	Encroachment Condition Module
EDMS	Environmental Data Management System
EFD	Engineering Field Division
E-GIS	Enterprise Geographic Information System
EMS	Environmental Management System
EOC	Emergency Operations Center
EOP	Emergency Operations Plan
EPA	Environmental Protection Agency
ERDC	Engineer Research and Development Center
ERP	Emergency Response Plan
ESA	Endangered Species Act
ESG	Environmental Support Group
ESOH	Environment, Safety, and Occupational Health
ESQD	explosive safety quantity distance
ESRI	Environmental Systems Research Institute
ESS	Explosive Safety Siting Tool
ESSENCE	Electronic Surveillance System for the Early Notification of Community-based Epidemics
EST	Eastern Standard Time
EU	European Union
EUCOM	European Command
FAA	Federal Aviation Administration
FEMA	Federal Emergency Management Agency
FHP&R	Force Health Protection & Readiness
FGDC	Federal Geographic Data Committee
FOS	forward operating site

FP	force protection
FS	Forest Service
FSA	Farm Service Agency
GAO	Government Accountability Office/General Accounting Office
GCSS	Global Combat Support System
GEOINT	Geospatial Intelligence
GI&S	Geospatial Information and Services
GIG	Global Information Grid
GIO	Geospatial Information Office
GIS	Geographic Information System
GISR	Geographic Information System Repository
GNOME	General NOAA Oil Modeling Environment
GPS	Global Positioning System
GRR	GeoReadiness Repository
HAF GIO	Headquarters Air Force Geo Integration Office
HAF/ILEI	Headquarters Air Force Geo Integration Office
HPAC	Hazard Prediction Assessment Capability
HSPD	Homeland Security Presidential Directive
I&E	installations and environment
ICRS	Integrated Range Control System
IEB	intelligence exploitation base
IEC	Infrastructure Executive Council
IED	improvised explosive device
IFS	Integrated Facility System
IFSAR	Interferometric Synthetic Aperture Radar
IGIR	Integrated Geographic Information Repository
IGI&S	Installation Geographic Information and Services
IMA	Installation Management Agency

IMCOM	Installation Management Command
INRMP	Integrated Natural Resources Management Plan
IPA	Intergovernmental Personnel Act
IPT	Integrated Product Team
IR	Installation Restoration
IRB	Investment Review Board
IRP	Installation Restoration Program
IRRIS	Intelligent Road and Rail Information Server Server
ISG	Infrastructure Steering Group
ISR	Installation Status Report
IT	information technology
ITAM	Integrated Training Area Management
ITIM	IT Investment Management
IVT	Installation Visualization Tool
JCSG	Joint Cross Service Group
JEPES	Joint Engineer Planning and Execution System
KORO	Korea Region Office
KWAN	Korea Wide Area Network
LEAM	Land-Use Evolution and Impact Assessment Model
LIDAR	Light Detection and Ranging
LUCA	land-use change analysis
MAJCOM	Major Command
MCAS	Marine Corps Air Station
MCB	Marine Corps Base
MCCS	Marine Corps Community Services
MEDCOM	Medical Command

MEGIN	Maryland Emergency Geographic Information Network
MGRS	military grid reference system
MILCON	Military Construction
MIM	Military Installation Map
MIMT	Military Installation Map template
MMD	mineral management division
MOA	memorandum of agreement
MOU	memorandum of understanding
MSA	metropolitan statistical area
MSDS	Mission Specific Data Standards
MTF	medical treatment facility
MTR	military training route
MWR	morale, welfare, and recreation
NAIP	National Aerial Imagery Program
NAS	Naval Air Station
NASA	National Aeronautics and Space Administration
NAVAIR	Naval Air Systems
NAVFAC	Naval Facilities
NAVSUPPAC	Naval Support Activity
NAWS	Naval Air Weapons Station
NCIS	Naval Criminal Investigative Service
NEPA	National Environmental Policy Act
NERMS	Navy Emergency Response Management System
NGA	National Geospatial-Intelligence Agency
NGO	nongovernmental organization
NGS	national grid system

NOAA	National Oceanic and Atmospheric Administration
NPS	National Park Service
NSA	Naval Support Activity
NTSB	National Transportation Safety Board
OIF	Operation Iraqi Freedom
O&M	Operations and maintenance
OMB	Office of Management and Budget
OSD	Office of the Secretary of Defense
OSHA	Occupational Safety and Health Administration
OUSD(AT&L)	Office of the Under Secretary of Defense for Acquisition, Technology, and Logistics
OASD(HD)	Office of the Assistant Secretary of Defense for Homeland Defense
OUSD	Office of the Under Secretary of Defense
PACAF	Pacific Air Forces
PDF	portable document format
PKI	Public Key Infrastructure
PM	program manager
POC	point of contact
POL	petroleum, oils, and lubricants
PRV	plant replacement value
QA	quality assurance
QAP	quality assurance plan
QC	quality control
R&D	research and development
RAB	Restoration Advisory Board
RAF	Royal Air Force
RCI	Residential Community Initiative

RCRA	Resource Conservation and Recovery Act
RCW	red-cockaded woodpecker
RDAP	Range Design and Planning
RDT&E	research, development, test, and evaluation
RMTK	range managers toolkit
ROI	return on investment
RPG	rocket propelled grenade
RPI	Real Property Inventory
RSIMS	Regional Shore Installation Management System
RSIP	regional shore infrastructure planning
SAF	Secretary of the Air Force
SAF/IEB	Assistant Secretary of the Air Force for Infrastructure, Environment, and Logistics/Basing and Infrastructure Analysis
SARNAM	Small Arms Range Noice Assessment Model
SARS	severe acute respiratory syndrome
SDDC	Surface Deployment and Distribution Command
SDDCTEA	Surface Deployment and Distribution Command Transportation Engineering Agency
SDI	spatial data infrastructure
SDSFIE	Spatial Data Standards for Facilities, Infrastructure, and Environment
SDZ	surface danger zone
SECDEF	Secretary of Defense
SERDP	Strategic Environmental Research and Development Program
SERPPAS	Southeast Regional Partnership for Planning and Sustainability

SIRRA	Sustainable Installation Regional Resource Assessment
SRA	Sustainable Range Awareness
SRP	Sustainable Range Program
SSIIS	Shore Station Integrated Information System
TABS	The Army Basing Study
T&ES	threatened and endangered species
TEC	Topographic Engineering Center
TIO	TEC Imagery Office
TMA	TRICARE Management Activity
TRANSCOM	Transportation Command
TSG	targeting support group or test signal generator
UAV	unmanned aerial vehicle
USACE	U.S. Army Corps of Engineers
USACHPPM	U.S. Army Center for Health Promotion and Preventive Medicine
USAF	U.S. Air Force
USAFE	U.S. Air Force Europe
USAREUR	U.S. Army Europe
USDA	U.S. Department of Agriculture
USFWS	U.S. Fish and Wildlife Service
USGS	U.S. Geological Survey
USMC	U.S. Marine Corps
USNORTHCOM	U.S. Northern Command
UST	underground storage tank
UTIS	Urban Tree Information System
UXO	unexploded ordnance
VA	vulnerability assessment
VMap	Vector Smart Map
W-EPS	Web Evacuation Permit System

WMD	weapons of mass destruction
W-Plan	Web Planning Tool
WPC	Warrior Preparation Center
W-Site	Web Siting Tool

CHAPTER ONE

Introduction

Background

U.S. military installations around the world, just like city, town, and county governments, use geospatial data assets for many purposes, including emergency response, environmental management, facility and infrastructure management and planning, homeland defense, public safety, and the provision of health care and other services. For example, geospatial data and associated attributed information about road and building locations and conditions and building functions are used to help dispatch appropriate fire and other resource vehicles during a fire or other emergency. These data are also used to help plan, manage, and maintain the community's buildings and roads.

Geospatial data assets, comprising digital databases, specialized software applications, documents, and web services that contain some type of geospatial information, are needed by many Department of Defense (DoD) echelons and agencies—from high-level policymakers to installation personnel. For example, the Office of the Secretary of Defense (OSD) uses geospatial data assets to help with strategic basing decisions, as in the 2005 U.S. Base Realignment and Closure (BRAC) process. At the installation, regional, and OSD levels, geospatial imagery data, such as high-resolution aerial or satellite images, are combined with other geospatial data in Geographic Information System (GIS) software[1] to help manage military training ranges and to help with installation land-use planning and natural resource management.

Geospatial data assets are used throughout DoD echelons and agencies to help with this business domain, which includes defense training, financial issues, force protection, installation, and environmental missions. Warfighting and intelligence missions are also supported by installations and environment (I&E) geospatial data assets, especially when combined with warfighting and intelligence geospatial data assets. For example, installation geospatial data about ports and runways are used in warfighting GIS-based systems to help plan deployments and logistical support to operations in Afghanistan and Iraq. I&E geospatial data assets are used by both the intelligence and installation communities in antiterrorism planning and analyses.

For many missions, sharing and integrating geospatial data assets can result in cost and performance efficiencies. For example, at a military installation, the Department of Public Works (DPW), the environmental management staff, and the training staff all need road data within their GIS software to help them perform their missions. These organizations can save on the cost of creating and maintaining road data by sharing. As an example, a shared and integrated GIS-based web system provides a common picture to personnel involved in homeland defense responses or bioterrorism, hurricane, or other emergencies. By having key geospatial information about the emergency integrated in one place and accessible by key staff, such as the installation commander, emergency responders, and utilities staff, preparedness and response are improved. Such integration enables a faster response time, helps decisionmakers manage assets better, and improves communications across the diverse agencies involved. In contrast, if data are not shared, organizations redundantly and wastefully develop and maintain the same datasets, which can result in inconsistencies and quality differences in the data. If one organization uses out-of-date or poor-quality data, the outcome of the decision or mission at hand can be affected. The benefits of sharing data assets accrue not just at local

---

[1] GIS is a class of software for managing, storing, manipulating, analyzing, visualizing, and using digital geospatial data.

installations but across installations, the Services, OSD, and other organizations.

Because these benefits are recognized, sharing such data has become DoD and U.S. federal government policy. The White House Office of Management and Budget (OMB) Circular A-16[2] and DoD Directive 8320.2[3] not only stress the need for coordinating, sharing, and integrating geospatial data assets across DoD and other federal agencies to improve the efficiency and effectiveness of agency activities, they require it. In July 2004, within the Deputy Under Secretary of Defense Installations and Environment Business Transformation (DUSD/I&E (Business Transformation)) directorate, a new organization, the Defense Installation Spatial Data Infrastructure (DISDI) Office, was created to help facilitate the sharing and integration of I&E geospatial data assets. The Business Transformation office wanted agencies to stop paying for redundant data that other agencies had already acquired.

The DISDI Office within DoD comprises people, policies, and practices necessary to acquire, steward, and share installation, environmental, and range geospatial data assets in support of defense, federal, and national goals. DISDI's first major initiative was to develop the Installation Visualization Tool (IVT) for the 2005 BRAC process.[4] The IVT was designed for "situational awareness" in that process and provided a way to view imagery and geospatial data in a consistent fashion for 354 sites, including ranges, meeting BRAC 2005 threshold criterion.

Given the information and geospatial technology advances of the past 10–20 years, it is now easier to use and share geospatial data assets. First, advances in the computer and broader information technology industry have helped bring down the cost of using geospatial systems and data. Data-processing and computing power have increased significantly and also have decreased in price. Such changes have made

---

[2] Office of Management and Budget (2002).

[3] Department of Defense (2004)

[4] The IVT program office started to develop IVT in 2003, then IVT was transformed into a task of the DISDI Office in July 2004.

the use of geospatial data and the processing of it—that is, investing in computer hardware and software—less expensive. Second, GIS software programs have evolved so that they have more capabilities and are easier and cheaper to use. Third, geospatial data are less expensive and simpler to acquire, create, and update. For example, Global Positioning System (GPS) has made it easier to enter accurate geospatial coordinates from the field. Fourth, advances in web geospatial applications and enterprise approaches have made it easier to share geospatial information across organizations. Fifth, the geospatial community has made significant strides in standards and specifications for geospatial interoperability and data sharing. A baseline of international and national consensus standards now exits to support the implementation of spatial data infrastructures (SDIs) at all levels. Given all these reasons, more military organizations, as well as other government organizations, are investing in their own geospatial data assets and the sharing of them.

However, barriers to sharing still exist. Because of the complex set of uses and organizations involved, it can be difficult to share and integrate such assets to maximize mission efficiency. Organizational stovepipe issues, such as installation environmental staff, DPW, and training staff not wanting to share their data, can limit geospatial data asset sharing. Organizational firewalls and security concerns also limit sharing. The fact that an upfront investment often needs to be made before the benefits are realized, not just in geospatial data and software development but in the sharing mechanisms, can limit investments in sharing. Furthermore, military organizations are not making the investments that would allow sharing geospatial data assets, because a business case has not been made for doing so, especially when it comes to sharing beyond a local installation. Also, there has been no rigorous analysis of the potential benefits.

Project Objective

Given this lack of rigorous analysis of the benefits of sharing geospatial data assets, RAND conducted a study for the DISDI Office on the mission effects of sharing geospatial data assets across the Global Information Grid (GIG). The objective of the study was to assess the net effects of sharing I&E geospatial data assets within the business domain and across the business, warfighting, and intelligence mission areas of the DoD GIG. Key tasks included

- analyzing key missions where I&E geospatial data assets are being shared or could be shared
- developing ways to assess the net mission effects of sharing
- estimating the net mission effects of current and potential future efforts to share such data assets
- recommending how the DISDI Office can help to maximize mission benefits by sharing.

Methodology

This research study was conducted over a 15-month period, from March 2005 through May 2006. Our assessment consisted of four integrated tasks:

1. a review of relevant literature
2. in-depth interviews of producers and consumers of geospatial data assets
3. an examination of sample geospatial data assets
4. the development and application of logic modeling and other evaluation approaches for assessing mission effects.

Our literature review covered relevant DoD, information technology, geospatial technology, and impact assessment literature. Relevant DoD literature examined included OSD, Service, and installation articles, documents, and web sites about policies, procedures, and examples about using and sharing geospatial data assets. The information technology literature was reviewed for assessments about sharing and the benefits of sharing data and software applications. The geospatial technology trade press was examined to identify articles about the sharing of and military uses of geospatial data assets. Broader functional journals, concerned with such topics as environmental management and health, were also searched for examples of geospatial applications, especially military ones. The impact assessment literature was searched for ways to evaluate mission effects. The literature review and interviews made it clear that logic modeling along with cost-benefit approaches would be most useful for our study, so we examined this literature in some depth.

Over 100 diverse DoD I&E geospatial data asset producers and consumers were interviewed for this study. Interviewees included mission geospatial data asset developers and users at Army, Navy, Air Force, and Marine Corps installations around the world. Service regional and Major Command (MAJCOM) geospatial data asset developers and users were also interviewed. We also interviewed facilitators for the sharing of geospatial data assets, including DISDI staff and the Service headquarters' Geospatial Information Office (GIO) directors and their staffs. Many of these staff may be also be users. Other DoD geospatial data asset developers, facilitators, users, and potential users, such as National Geospatial-Intelligence Agency (NGA) staff, were also interviewed. In addition, a small sample of other federal agency, state and local government, nongovernmental, industry, and university geospatial data developers or users were interviewed, where relevant.

Most of the interviews lasted from an hour to two hours although some were as short as 20 minutes or as long as three hours. For the majority of interviews, standard questions were asked about how the interviewee's organization developed, used, could use, or shared geospatial data assets; the mission effects of such use and sharing; the barriers to sharing; and suggestions for what could be done by the DISDI Office to address the barriers and facilitate more sharing to improve mission operations.

We also spent a day interviewing both geospatial data asset developers and users at one installation for each Service: the U.S. Army's Aberdeen Proving Ground (APG), Maryland; U.S. Marine Corps (USMC) Camp Butler, Okinawa, Japan; Langley Air Force Base (AFB), Virginia; and Naval Air Station (NAS) Patuxent River, Maryland.

These installations were chosen because they are technology innovators and models for the use and sharing of geospatial data assets.

We examined a range of I&E geospatial data assets—from samples of hard copy maps, portable document format (PDF) map files, and Microsoft PowerPoint slides that were used in different mission applications to the web viewers designed for viewing installation geospatial data, such as the Army's Geographic Information System Repository (GISR). Diverse geospatial datasets were acquired and examined from a sample Army and U.S. Air Force (USAF) installation and from the IVT that was used in the BRAC process.

To assess mission effects, we examined a range of approaches and determined that information flow models, logic modeling, and, where feasible, benefit-cost analysis combined with the logic model were the most appropriate tools to use. We then developed and applied a methodology for assessing the mission effects that used information flow models and logic modeling combined with the benefit-cost analysis approaches. This assessment methodology is explained in Chapter Seven.

Organization of the Report

The next chapter explains which I&E geospatial data assets are being shared, who is sharing and why, and how the data are shared. Chapter Three discusses the many diverse ways that I&E geospatial data assets support traditional installation and OSD business mission areas and Chapter Four discusses the application of I&E geospatial data assets to warfighting operations. Because it is an important cross-departmental case study of data sharing, with far-reaching implications throughout DoD and beyond its original purpose, the development and uses of IVT data, both to support the BRAC process and external to BRAC, are discussed in Chapter Five. Chapter Six discusses likely trends in the future use and sharing of I&E geospatial data assets and some of the barriers that need to be surmounted to increase such sharing. Chapter Seven presents our assessment of the mission effects of using and sharing I&E geospatial data assets. This chapter describes a methodology and approach that DISDI and other organizations can use to assess the affects of geospatial data use and sharing. The final chapter presents the conclusions from this study and recommendations for the DISDI Office to help facilitate more sharing of I&E geospatial data assets. The appendix presents, by mission area, over 130 examples of how I&E geospatial data assets enable business-related missions. This appendix is especially useful to readers interested in a specific mission area, such as homeland defense, environmental management, or military health.

CHAPTER TWO

What Is Shared, Who Is Sharing It, Why, and How

This chapter describes the diverse set of installation and environment geospatial data assets that are currently being shared or could be shared, who develops and maintains such assets, who is sharing the assets with whom, for what purposes, and how. It is important to note that even though there is currently a large amount of sharing, some barriers still exist. These barriers are discussed in Chapter Six.

Diverse I&E Geospatial Data Assets Are Being Shared or Could Be Shared

Military installations develop, maintain, and use a diverse set of installation, environment, and range geospatial data assets to conduct installation business. By definition, these I&E geospatial data assets are for permanent installations contained in the official OSD Real Property Inventory (RPI).[1] Such assets are not associated with forward operating locations or other temporary contingency operation support installations, which have their own geospatial data assets. Geospatial data assets to support such warfighting operations are developed and maintained by warfighting organizations within the DoD. NGA has the primary mission to provide Geospatial Intelligence to the warfighting and intelligence communities. DoD Directive 5105.60, which created this agency in 1996, states that NGA's mission is to "provide timely,

---

[1] This definition—that I&E geospatial data assets refer to permanent installations—is important to remember. Throughout the rest of this monograph, when the term "installation" is used, it refers to permanent installations unless otherwise noted.

relevant, and accurate imagery, imagery intelligence, and geospatial information in support of the national security objectives of the United States."[2]

U.S. Army, USAF, U.S. Navy, and USMC installations across the world are developing, using, and sharing I&E geospatial data assets. Such assets include digital geospatial data, software applications that use geospatial data, and other products that use geospatial data, such as hard copy maps.

Digital Geospatial Data

One of the most common and fundamental types of installation geospatial data assets is GIS datasets. GIS is a class of software for managing, storing, manipulating, analyzing, visualizing, and using digital geospatial data. Within these software systems are different digital geospatial data layers and imagery about installations. Many installations develop and maintain hundreds of GIS data layers. Datasets at different levels of scale and time periods are often maintained because of different needs.

Table 2.1 shows sample GIS data layers for USMC Camp Lejeune to illustrate the wide range of data being developed, maintained, used, and potentially available for sharing. Camp Lejeune calls its GIS the Command Integrated Geographic Information Repository (IGIR) and it contains hundreds of GIS datasets, both vector and imagery data, and they are shared and used by numerous organizations across the installation. Some sample categories of GIS data and sample datasets are presented in this table. Besides these data, the installation IGIR data also include information about the region, such as county boundaries, land-ownership data, hazardous waste disposal sites, regional hospitals, watersheds, and coastal reserve areas.

In each Service, some fundamental data layers are so widely used and needed for viewing and understanding an installation that they are considered and called Common Installation Picture (CIP) data layers. Such basic data layers are designed to be used and shared

---

[2]   Department of Defense (1996, p. 2).

Table 2.1
Sample GIS Data Layers for Camp Lejeune

Category of Dataset	Sample GIS Datasets
Boundary	Camp Lejeune installation boundaries Emergency medical zones Areas of base responsibilities
Buildings	Existing structures Floor plans Tower structures
Cartographic	Base mapbook reference grids Existing conditions county boundaries Military Installation Map (MIM) image border and grid points
Climate	Sensors and weather stations
Communications	Communication antennas Telephone fiber-optic and copper cables Telephone manholes
Cultural	Monuments and markers
Environmental hazards (combines characterization of hazards, hazardous waste, pollution control, solid waste, and some other subcategories)	Water sampling sites Groundwater monitoring sites Hazardous material storage Air-quality emission points Groundwater pollution plumes Aboveground and underground storage tanks (USTs) Leaking USTs Installation Restoration (IR) Comprehensive Environmental Response, Compensation and Liability Act (CERCLA) sites

Table 2.1—continued

Category of Dataset	Sample GIS Datasets
Fauna	Closed landfill gas and monitoring wells Compost facilities Solid waste management units Recycle sites
	Red-cockaded woodpecker (RCW) habitat management area, cavity trees, foraging areas, and recruitment sites Shellfish sampling sites Nesting areas and sites (non-RCW) Wildlife openings Bird Aviation Safety Hazard (BASH) data
Flora	Fire breaks Prescribed burn/wildfire history Locations of threatened and endangered plant species and buffers for them Timber management stands and compartments Plant communities
Hydrography	Ditches Creeks and streams Water bodies Wetland areas
Improvement areas (combines general and recreation subcategories)	Formal base access gates Flag poles Miscellaneous improvement structures Athletic field areas Recreation boat ramps

Table 2.1—continued

Category of Dataset	Sample GIS Datasets
	Golf course fairway and cart paths
	Hunting areas
	Playgrounds
	Outdoor recreation areas
	Recreation trails
	Outdoor swimming pools
Land status	Cemeteries
	Grounds maintenance (mowing)
	Land covers
	Land-use areas
Military operations	Ammunition storage areas
	Landing and drop zones
	Firing area, lanes, and points
	Special use airspace
	Military gates and observation towers
	Live fire range fans
	Range footprints
	Amphibious drop zones
	Military targets
	Impact areas
	Tanks pads and trails
	Military training areas

Table 2.1—continued

Category of Dataset	Sample GIS Datasets
Transportation (combines air, pedestrian, railroad, vehicles, and water subcategories)	Airfield surfaces Pedestrian footbridges and sidewalks Railroad bridges and centerlines Road edge, centerlines, bridges, and signs Driveways Parking lots Docks and piers
Utilities (combines electrical, fuel, gas, storm, wastewater, and water subcategories)	Electrical cable group, generator sites, and manholes Electrical regulators and substations Exterior lights Electrical poles Gasoline pumps Heating system boilers, fittings, and manholes Heating system lines and valves Storm sewer basins, headwalls, and inlets Storm sewer lines/pipes and pipe ends Storm sewer pond risers Wastewater manholes, lines, and pump areas Wastewater pumping stations and treatment plants Wastewater septic tank points Water distribution zones Water lines, meters, tanks, and supply wells Water treatment plants Fire hydrants

Table 2.1—continued

Category of Dataset	Sample GIS Datasets
Imagery	LandSat satellite imagery, October 1993, 1996, and 1999
	1993–1996 vegetation changes
	IKONOS satellite imagery, 2000–2004
	2004 natural color digital orthophotography (basewide, 30 cm and built-up areas, 15 cm)
	Surface digital elevation model (DEM), 5 ft and 20 ft
	Historical aerial imagery, 1930s–1980s
	Camp Lejeune MIM
	U.S. Geological Survey 1:24K, 1:100K, and 1:250K quadrangles
	Building photographs
	2005 3-D fly-through virtual range tours

SOURCE: Camp Lejeune (2005).

NOTES: These data have been organized slightly differently to present them here and some categories were left out. For example, the "common" category was left out. However, dataset names were taken directly from the original source. For a full description of Camp Lejeune datasets, see the table source.

by organizations across an installation so that one "map" or one set of geospatial data themes can be shared. Historically, at many installations, different functional organizations, such as facialities management, training, and environmental, would develop and produce their own GIS datasets, which was not cost-effective. When different organizations develop and maintain the same data, inconsistencies and quality differences in the data result. If one part of the installation uses out-of-date or poor-quality data, it can affect the outcome of an operation using those data, especially if coordination with other organizations is required. For these reasons, installations developed the idea of the CIP to provide the same geospatial view across the installation.

Different military Services have defined different CIP data. Table 2.2 shows the CIP for the Air Force and Table 2.3 for the Army. The Navy is taking a regional approach and has developed its own regional version of the CIP, i.e., geospatial datasets that are a minimum requirement for all regions. Similarly, the Marine Corps has developed its own version of CIP data that every installation should maintain.

The USAF effort developed the CIP to be the common installation GIS datasets that provide the basic foundation for base management needs. The 2006 CIP consists of 36 vector and one raster data layers.

The U.S. Army has also defined its own set of most basic data layers for each installation. The U.S. Army's CIP GIS data layers are shown in Table 2.3. These data include many of the same categories of data used by the USAF; however, because of the importance of ground testing and training ranges and ground training to the Army, its CIP includes a great deal of data about ranges and other training areas, whereas the USAF CIP includes more information related to runways and aircraft.

Each Service has also created specialized data layers for individual mission functions, often called mission-specific datasets. Table 2.1, which showed sample GIS datasets for Camp Lejeune, illustrated some of the fauna and flora categories used to manage natural resources, such as the longleaf pine forests. However, even though such datasets are developed for a specific mission function, they could be used or shared by another mission and for other functions in that mission area.

Table 2.2
USAF 2006 CIP Data Layers

Category of Dataset	Data Layer
Buildings	Tower areas Slab areas Structure existing sites
Cadastre[a]	Installation areas
Hydrography	Shorelines Surface water body areas Surface water course areas Surface water course centerlines
Improvement	Athletic court areas Athletic field areas Golf course areas Swimming pool areas Dam sites Levee berm areas Fence lines Gate lines Wall lines Campground areas Playground areas Recreation park areas Recreation trail centerlines
Landform	Elevation contour lines
Military operations	Military range areas
Transportation	Airfield surface centerlines Airfield surface area Tunnel areas Footbridge areas Pedestrian sidewalk areas Railroad bridge centerlines Railroad centerlines Road bridge areas Road bridge centerlines Road centerlines Road areas Vehicle driveway areas Vehicle parking areas
Imagery	1-meter resolution imagery—the minimum required for the cantonment area; lower-resolution imagery of unimproved or nonbuilt areas of the installation

SOURCE: Headquarters Air Force Geo Integration Office (2006).

[a] "Cadastre" refers to mapping boundaries of some type, defined as "the man-made division of land into areas of ownership and control" (Headquarters Air Force Geo Integration Office, 2006, p. 9).

Table 2.3
U.S. Army CIP GIS Data Layers

Category of Dataset	Data Layer
Boundary	Political jurisdiction county areas Political jurisdiction municipal areas Political jurisdiction state areas
Buildings	Buildings
Cadastre	Installation boundaries Parcels
Communications	Antennas
Environmental hazards	Hazardous materials storage sites
Flora	Land vegetation
Geodetic	Control points U.S. Geological Survey (USGS) quad areas
Hydrography	Surface water course areas Surface water course centerlines Wetlands
Improvement	Gates
Landform	Elevation contour lines Spot elevation points
Military operations	Military special use airspace areas Ammunition storage areas Firing sites Forward arming and refueling points Dudded impact sites Non-dudded impact areas Military training subareas Military restricted access areas Tank trail lines Military drop zone areas Military landing zone areas Military live fire range areas Military observation points Military range areas Military surface danger zone areas Military surface danger zone lines Training sites Operational range areas
Transportation	Airfield surface sites Railroad centerlines Road centerlines

Table 2.3—continued

Category of Dataset	Data Layer
Utilities	Electric cable lines
	Power substations
	Pipelines
Miscellaneous	Military landing zone points

SOURCE: Data courtesy of the U.S. Army Geospatial Information Office, Fall 2006.

For example, a forest species geospatial data layer at Camp Lejeune, Fort Bragg, Fort Benning, or Eglin AFB used to help manage longleaf pine forests could also be used to help manage the endangered RCW and other T&ES, plan and assess controlled burns, analyze encroachment, develop environmental impact assessments, respond to firefighting emergencies, and manage training ranges.

Many installations in each Service have developed and are using such CIP and mission geospatial data. However, many have not yet fully developed their data. To better understand geospatial data capabilities, each Service conducted an installation inventory during the spring and summer of 2005. We briefly present some sample survey results to show that even though installations have done a lot to develop, use, and share I&E geospatial data assets, the Services still have a lot to do to fully develop this capability for supporting different missions.

DISDI worked with the Service GIOs to develop some common questions for the survey and to define some basic geospatial datasets for different mission areas, including DoD Critical Infrastructure Protection, Real Property Inventory (RPI), and Environment, Safety, and Occupational Health (ESOH). Questions about these mission areas were used to help quantify the amount of mission capability present at each installation. At the time, the USAF found that 70 percent of the 2005 CIP layers were available across the USAF. However, it found that the program could support, on average, only 27 percent of the essential mission datasets for the ESOH, 56 percent for the DoD Critical Infrastructure Protection, and 60 percent for the RPI missions.[3]

---

[3] Headquarters Air Force Geo Integration Office (2005b, p. 18).

.

Other Service surveys yielded similar or even less-advanced program results.

Digital geospatial data can also include other types of geospatial data that are entered and stored in computer systems, for example, GPS coordinates and associated attribute data that are entered in a hand-held device from the field, such as fire hydrant locations. Similarly, an Excel spreadsheet or ACCESS database that contains data by address, such as the home addresses of military dependents, constitutes geospatial data.

Software Applications That Use Geospatial Data

Customized GIS software applications, both web-based and non-web-based, are developed so that users can view, share, and use the geospatial data for many different purposes. Such software are developed and applied for general viewing of geospatial information and for specific applications for a functional mission, such as emergency response and training range management tools.

Commercial examples of web-based geospatial data assets for general viewing of geospatial data are Google Earth and Google Maps. An example of a web-based system for viewing geospatial data at a military installation is at NAS Patuxent River, where users can access different I&E geospatial data assets from the web for diverse business functions. Users include morale, welfare, and recreation (MWR), public works, base administration, and ESOH staff. The system is part of the Naval Air Systems (NAVAIR) Command regional shore infrastructure planning (RSIP) approach that focuses on providing operating force support, community support, and base support. A sample of a web-based system for viewing geospatial data across different Service installations around the world is the DISDI Viewer, shown in Figure 2.1.

The DISDI Office has developed this geospatial viewer so that organizations across DoD can view general installation geospatial data (installation point locations, site boundaries, imagery, wetlands, etc.) overlaid on national- and regional-scale geospatial data, providing situational awareness and enabling strategic decisionmaking activities; perform queries of installation and related data; and make simple maps for printing or embedding in presentations.

Figure 2.1
The DISDI Viewer

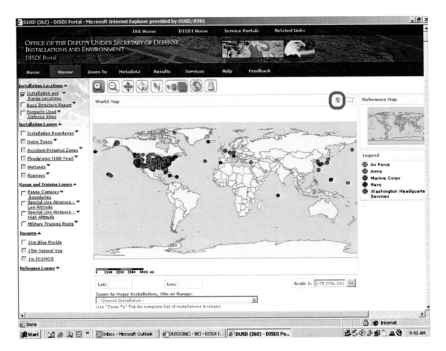

SOURCE: Image courtesy of the DISDI Office, 2006.
RAND MG552-2.1

Such web-based systems will be discussed more below, because they are also key methods for disseminating and sharing geospatial data with many people and are an important growing trend for future data sharing.

Also, many I&E geospatial data software applications that are not web-based are currently being developed and used. For example, in June 2004, the USMC conducted a survey of the number of desktop GIS tools being used at its installations and found 185 being used at 11 installations.[4] I&E geospatial data software applications range from general GIS-based toolsets to simple and sophisticated mission-specific applications. A general GIS-based toolset example is the

_____

[4] URS Group, Inc. (2004, p. 5).

USMC's GEOFidelis,[5] a desktop GIS toolset to be implemented throughout the Marine Corps. These tools include general geospatial viewing and analysis tools, such as specialized zoom, sum, viewing, printing, and exporting tools. They also include a set of antiterrorist/ force protection tools, which is an example of a toolset for a specific mission function.

Such software tools may be accessible through downloads from a web site or run through a web site, but they are not stand-alone web-based systems that can be run from any computer. For example, at Ramstein Air Base (AB), Germany, installation users with a need to know can download an Asbestos Management System software application from the installation's GIS-based web service for examining asbestos concerns during building renovation work orders. The downloaded application software allows users to run this system on their desktops and to access and transfer data through the installation GIS-based web service to examine where asbestos is located in a particular building on the base and to notify others about it.

A wide range of these customized I&E GIS software applications, both web-based and non-web-based, are currently being used throughout different parts of DoD. Additional examples of such GIS-based software application tools are shown in Table 2.4.

Other Products That Use I&E Geospatial Data

Geospatial data assets also include other products using geospatial data, such as documents, hard copy maps, and videos. For example, warfighters use MIMs—installation maps employed in range training exercises, and drivers use hard copy installation road maps to determine how to drive to locations on an installation. Training videos that use the actual imagery for an installation are another example of a geospatial data asset developed for a specific mission.

---

[5]   In 2004, Marine Corps headquarters launched an initiative, known as the Marine Corps Installations Spatial Data Infrastructure, or "GEOFidelis," to standardize GIS within the Service.

Table 2.4
Sample I&E Geospatial Software Tools and Other Applications

Application Name	Short Description	Sample Use	Sample Data Used
Antiterrorism (AT)/ force protection (FP) tool	Marine Corps Base Camp Pendleton tool for AT/FP planning, analysis, and management	Helps assess possible AT/FP event, such as setting/ removing barriers and perimeters for buildings	Buildings, roads, sidewalks, barriers, and other installation assets
Assessment System for Hazardous Surveys (ASHS)	Tool to assess capacities for explosive safety for explosive hazards reduction	Assesses explosive safety at Royal Air Force (RAF) Welford, United Kingdom	Buildings, roads, railways, munitions storage areas, and imagery
Blast Noise Prediction (BNOISE) computer model	U.S. Army Corps of Engineers (USACE) model that estimates blast noise exposure from ground training	Estimates ground training range noise exposure level values for both humans and animals	Training ranges, buildings, imagery, and basic flora and fauna information
Consequence Assessment Tool Set (CATS)	Consequence management tool that employs a suite of hazard models to estimate and analyze effects from natural and manmade disasters	Used by installations and other organizations to help assess, train, and plan against potential terrorist events as well as natural disasters	Installation boundaries, buildings, roads, waterways, weather, and wind patterns
Emergency Response Tool Suite	Laptop emergency response planning tool for Ramstein AB	Used during incidents and training exercises to help with emergency response operations and evacuation planning	Installation boundaries, roads, buildings, key personnel contact information, runways, and water features
Facility utilization tool	Langley AFB tool to help manage base building and other facility utilization	Used to help with facility management, maintenance, and contractor oversight by examining floor plan details and work order requests	Buildings, floor plans, and work order information

Table 2.4—continued

Application Name	Short Description	Sample Use	Sample Data Used
IVT	Desktop GIS Viewer to view geospatial data and imagery for the 2005 BRAC process	Provides situational awareness to BRAC decisionmakers	Installation and range boundaries, noise contours, accident potential zones, wetlands, and imagery
Military health system atlas	OSD atlas of military medical capabilities and military populations	Used to help with military treatment facility resource decisions	States, counties, cities, installation boundaries, military treatment facilities, and military populations
National Environmental Policy Act (NEPA) checklist	NAS Patuxent River GIS-based tool to document the NEPA process	Used to determine and document what must be done as part of the NEPA process	Buildings, roads, wetlands, storm water management areas, and T&ES and cultural resource information
Range managers toolkit (RMTK)	A GIS-based toolset for analyzing and developing ground training ranges	Used by Army and USMC bases to help locate a new training range and to assess safety, noise, and other effects	Roads, railroads, training areas, buildings, imagery, and water features
U.S. Army Europe (USAREUR) Integrated Training Area Management (ITAM) Viewer	2-DVD set to view USAREUR training areas on desktop, includes zoom, measuring tools, printing, and exporting to PowerPoint	Used to view training lands and help plan training exercises before arriving at an installation	Aerial imagery, topographic maps, and vector data for more that 80 USAREUR training areas
Web-based installation general plan	RAF Mildenhall tool with live link and map to the installation general plan	Used by installation planners to help with base planning	Base boundaries, buildings, roads, streams, airfields, runways, and other asset datasets

Who Creates, Maintains, and Updates I&E Geospatial Data Assets

Before discussing who shares these assets, it is important to understand who creates, maintains, and updates them. Most of the Services' basic digital geospatial data, i.e., the CIP and mission GIS datasets, are created, updated, and maintained at the installation or regional level. Responsibility for these data differs from Service to Service and even from installation to installation. Historically, the mission functional staff that needed the data created, maintained, and updated them; for example, environmental staff developed the environmental GIS layers, DPW staff developed building and roads infrastructure data, and range staff developed training range datasets. The mission experts are considered the mission data stewards because they understand, use, and know their mission data the best.

In the last five to ten years, more installations and even some Services have taken a more centralized approach to developing and maintaining basic geospatial data, because of the benefits of sharing and having a more centralized approach, coupled with the advances in enterprise software systems. The main advantages to centralized approaches include the following: Data are not being re-created, which saves time and money and helps avoid problems with creating and accessing inconsistent versions of the same data; and it is easier to share the data and make them available to a wide range of users. The main disadvantage to such approaches is that they require an investment in a centralized data capability, which often is difficult to secure and maintain resources for, i.e., it often is easier to acquire funds to invest in geospatial data to support a specific application task, because resources are allocated along business lines and cross-business capabilities are very difficult to fund.

Enterprise-based computer systems appeared in the late 1990s to help large organizations manage distributed information resources across their entire organizations. Such systems initially focused on office automation, enterprise-wide resource planning, and document management. Only in the last few years have enterprise systems been developed for the management, use, and sharing of geospatial information. For example, Camp Lejeune's IGIR is an enterprise GIS designed as a centrally managed database containing hundreds of layers of geographic information, imagery sets, engineering drawings, photographs, and links to external databases. Many organizations at the installation can access and use this geospatial information from this central point of access.

For installations and Services that take a more centralized approach, such as Camp Lejeune, the CIP data are often developed and maintained by a GIS function or shop. This GIS shop may also take over responsibility for maintaining some of the mission-specific datasets. However, the GIS staff often still rely on the mission experts to supply them or help them update these datasets. Function staff and GIS shops also develop and maintain other types of geospatial data assets, such as application programs.

Each Service and individual installation has a different level of centralization in its development of I&E geospatial data assets. The U.S. Army has the most decentralized program at the installation level, where installations have staff in environmental, DPW, and range departments, as well as in other areas, such as information technology departments, who develop, maintain, and update geospatial data assets. A few Army installations, such as Aberdeen Proving Ground, have taken a more centralized approach by creating a GIS shop that supports the installation. The Army is trying to develop a lead single point of contact for installation geospatial information by establishing a geospatial coordinator at each installation. The USAF has already centralized its geospatial data functions at each installation under a program called GeoBase, where every installation has an officer responsible for geospatial data assets. The U.S. Navy has taken a regional approach, where different Navy regions have a GIS shop responsible for developing and maintaining geospatial data assets. However, individual Navy installations, especially larger ones, also develop and maintain geospatial data. In 2006, the USMC also began to take a regional approach, although most USMC installations still develop and maintain their own data.

In addition to the installation staff developing and maintaining geospatial data assets, Service headquarters, functional commands, and regions also develop, maintain, and update geospatial data assets.

Other DoD staff, such as DISDI and NGA, also create and fund the creation of I&E geospatial data assets, such as DISDI creating the DISDI Portal. OSD organizations, however, focus more on software applications and rely on the installations to supply them with the basic I&E geospatial datasets for those applications. OSD organizations may also generate strategic geospatial datasets, especially ones designed for looking across a region, the nation, or the world, such as a georeferenced point dataset showing worldwide installation locations.

Who Shares and Who Uses I&E Geospatial Data Assets

I&E geospatial data assets are used and shared at many different organizations and levels within and outside DoD. Organizations and individuals act as I&E geospatial data asset developers, users, and facilitators. A facilitator is someone who helps promote and enable the development, use, and sharing of the assets.

Here we discuss which organizations are sharing data within and across different levels of DoD. We discuss how data are shared across different organizations at an installation, within different levels of a military Service, in other parts of DoD, and outside DoD.

Sharing Across Different Organizations/Mission Functions at an Installation

First, geospatial data assets are shared across different organizations and mission functions at an individual installation. For example, Ramstein AB, Germany, has a web-based system for viewing geospatial information for the base that is used by over 1,000 users[6] from diverse organizations around the installation. Ramstein's web system users include

---

[6] This statistic is from August 2005 and includes active users. One-time users are not included in this user count. Ramstein tracks who logs on, how many times, and for how long. In fact, if people have used the system only once or twice, an email is sent asking if they still want access. If they do not, they are taken off the system. By May 2006, use declined to less than half this number of users, probably because half the users were then using the Air Force portal to view Ramstein maps for more routine applications, such as looking up the locations of buildings. Despite the change in numbers, the distribution of active system users is about the same.

commanders, civil engineers from public works, environmental staff, communications staff, the Combat Support Wing, security forces, the intelligence community, first responders, the housing office staff, airfield operations staff, warfighters, and medical personnel. Figure 2.2 shows the number of Ramstein's geospatial web service users by main organizational areas.

Figure 2.2
Ramstein AB Geospatial Web Service Users

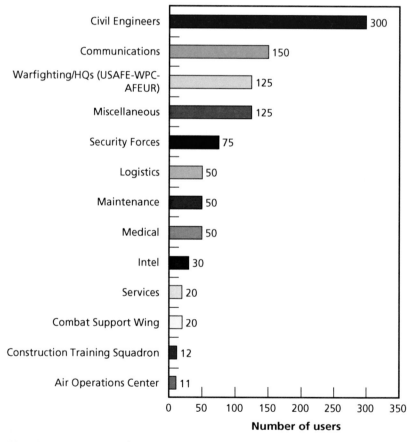

SOURCE: Data courtesy of Ramstein Air Base, 2005.
RAND *MG552-2.2*

The largest group of users includes civil engineers, who include the environmental and installation construction and planning staffs. It is important to note that this figure shows the number of users, not the functional use of the system. For example, it does not tell if intelligence staff members are using the system for intelligence purposes or to provide directional maps for the base.

Sharing Across Different Levels Within a Military Service

Second, geospatial data assets are shared among different levels of an individual Service, from a regional level, to functional commands and headquarters level.

Each U.S. military Service develops, shares, and uses geospatial data assets at a regional level or by functional commands. Since the U.S. Navy has taken a regional approach to the development and use of its installation geospatial data and information support, each Navy region has an office for providing geospatial data services. For example, the Navy Engineering Field Division (EFD) South is the regional office that provides GIS services for the 26 Navy installations in states in the Southeast and near the Gulf Coast, from Texas to Florida and up to South Carolina. USAF geospatial data support, on the other hand, is funded and organized by Major Commands, such as Air Combat Command (ACC), Air Mobility Command (AMC), and U.S. Air Force Europe (USAFE).

At this regional or functional command level, geospatial information is often used to help manage assets or analyze issues across the region or functional organization. For example, staff members with Navy EFD South have used their installation geospatial data to help manage buildings and other assets across multiple installations to achieve economies of scale in maintenance contracting. Other Service functional organizations below the Major Commands, such as the U.S. Army Center for Health Promotion and Preventive Medicine (USA-CHPPM) and other medical organizations within the Army Medical Command (MEDCOM), also use I&E geospatial data assets to support their mission operations (for examples, see the appendix).

Each Service has headquarters geospatial organizations to facilitate the development, sharing, and use of geospatial data assets. They facilitate the sharing of geospatial data assets within their respective Services by setting policies, by being a Service point of contact (POC) for geospatial data requests (that they usually forward onto the appropriate Service organization), and by sponsoring the development of Service-wide geospatial data web viewers so that many different military users can access I&E geospatial data assets. Each office also participates with the DISDI Office to establish a DoD-wide GIS community. Most of these organizations are also geospatial data users and provide specialized geospatial data support to other Service headquarters organizations. Table 2.5 provides the office name for each of these organizations.

These Service organizations essentially are developing their own spatial data infrastructures (SDIs). An SDI encompasses policies, standards, and procedures for organizations to cooperatively produce and share geographic data. Components of an SDI usually include institutional arrangements, policies and standards, data networks, technology

Table 2.5
Service Headquarters Organizations Responsible for I&E Geospatial Data Assets

Military Service	Organization Responsible for I&E Geospatial Data Assets	Program Name	Organization's Web Site
U.S. Air Force	Headquarters Air Force Geo Integration Office (HAF GIO)	U.S. Air Force GeoBase	https://www.my.af.mil
U.S. Army	Office of the Assistant Chief of Staff for Installation Management (ACSIM)	Army Installation Geographic Information and Services (IGI&S) program	https://gis.hqda.pentagon.mil
U.S. Marine Corps	Headquarters Marine Corps Installations and Logistics Department	GEOFidelis	http://www.geofidelis.net
U.S. Navy	Naval Facilities Engineering Command, Base Development	GeoReadiness	None

users, data, databases, and metadata.[7] A geospatial data clearinghouse acquires, maintains, and distributes geospatial data or provides informational services about data to many different data users. Such an organization may also integrate the data, generate the data, or perform other types of data-processing functions. Often, an SDI effort includes the establishment of a geospatial data clearinghouse to help advertise and provide access to data stores. Each Service also oversees the development of geospatial data clearinghouses in the form of Service geospatial data repositories.

Other Service headquarter organizations also are geospatial data users and developers. For example, USAF headquarters, Air and Space Operations, uses I&E geospatial information to help manage air and space operations.

Sharing with Other Parts of DoD, Including DISDI's Facilitator Role

Geospatial data assets are also shared across different Services and other DoD organizations for such mission functions as joint facility and environmental management, joint training, warfighting, and intelligence. Regional environmental concerns, such as for watershed or ecosystem management, are common areas for cross-service geospatial data sharing at the local and regional levels. In the 2005 BRAC process, I&E geospatial data from all military Services were used by OSD and the military Services to help assess BRAC options. For example, IVT data were used by the education and training BRAC Joint Cross Service Group (JCSG) to help assess joint training options.

The OSD DISDI Office is a facilitator for sharing I&E geospatial data assets across DoD. The creators of DISDI envision an institutionalized process where installation geospatial data (including data in GIS, computer-aided design and drafting (CADD), and imagery formats) are assembled, disseminated, and maintained in a fashion that supports DoD installation management and strategic basing decision missions worldwide. DISDI focuses on the business processes, people,

---

[7] Metadata are data about the geospatial data, including information about the content, quality, condition, and creators of the dataset.

and policies needed to provide visualization and mapping capabilities at installations. DISDI is not an information technology (IT) system but rather a mechanism by which geospatial data stewarded at and by DoD installations can be shared with stakeholders to meet their requirements. Since it was formally created in July 2004, the DISDI Office has undertaken a number of efforts to facilitate the development, use, and sharing of I&E geospatial data assets across the GIG.

First, DISDI completed development of the IVT for the 2005 BRAC process. Second, DISDI developed a web-based geospatial data portal for viewing such data, as shown in Figure 2.1, and has been a user of geospatial data assets supporting other OSD offices, such as the ADUSD ESOH. The DISDI Portal enables DoD organizations to learn about the availability and fitness-of-use of DoD installation and environmental geospatial data, view strategic maps of defense installations and surrounding regions, and learn where to access and obtain geospatial data from the Services for local visualization and analysis purposes. Third, DISDI created the DISDI community of interest (CoI), which has been registered with the DoD Chief Information Officer (CIO) as a cross-functional, institutionalized venue where those DoD missions with an interest in better leveraging I&E geospatial information resources can share perspectives and learn of emerging architectures, requirements, policies, and related initiatives. Fourth, holding and encouraging national, regional, and local DISDI conferences and participating in mission functional and geospatial conferences have been a part of DISDI's activities to help develop the CoI. Fifth, DISDI has been working on several projects to develop, test, and improve collaboration with other organizations, such as NGA, the Department of Homeland Security (DHS), and state and local governments. These projects include new collaboration between installations and NGA to develop Military Installation Maps to support the warfighter, and geospatial data-sharing models for the Project Homeland Colorado Pilot interagency effort to use geospatial assets in homeland defense and security activities. Both of these examples are discussed in later chapters. Sixth, DISDI has provided resource support to the USACE Topographic Engineering Center (TEC) Imagery Office to help the I&E community acquire commercial satellite imagery and save costs. Last, DISDI has also been providing technical assistance to various OSD offices on such issues as real property and environmental management.

Another example of how I&E geospatial data assets are shared within DoD can be seen in the activities of the NGA. NGA provides timely, relevant, and accurate Geospatial Intelligence support of national security by supporting the warfighter and intelligence communities within DoD, such as supporting U.S. homeland defense. NGA has used I&E geospatial data assets to develop MIMs for military training and in homeland security activities, such as developing Palanterra, a web-viewing system for U.S. geospatial information to support homeland defense and security. In addition, effective April 27, 2006, NGA was formally identified to OMB as the DoD senior agency for managing geospatial information.

Sharing with Organizations Outside DoD

Our study focused on the sharing of I&E geospatial data assets across DoD. However, we also found a large amount of current and potential future sharing outside DoD—among other federal agencies and state and local governments. These other U.S. government agencies also need geospatial information to help with key government functions, such as homeland security, natural resource management, and disaster preparedness and response. Often, the military works with other agencies in such functions and collaboration is increasing in these mission areas. For instance, geospatial data assets are shared with DHS and other government agencies to help in homeland security planning and training exercises. Installation map products are shared routinely with the Environmental Protection Agency (EPA) and state and local environmental regulators in environmental compliance reports and documents, such as NEPA reports. At the local level, military installations share their I&E geospatial data assets with local governments to help with joint infrastructure, transportation, utility, and natural resource management and for emergency planning and response. For example, Fort Bragg, Pope AFB, and the City of Fayetteville, North Carolina, have shared data about utilities and drainage interfaces. Sharing installation geospatial information has also been useful to Fayetteville in developing a hurricane evacuation plan.

Other DoD organizations also depend on other government agency and even industry data, such as utility company data. For example, OSD conducts critical infrastructure vulnerability assessments at installations across the world. It needs geospatial information from each installation, such as utility information, to help assess where vulnerabilities exist in base critical infrastructure.

Besides sharing with different parts of the U.S. government and the commercial sector, DoD organizations need to share with universities, nongovernmental organizations (NGOs), and allied governments. Many installations already share geospatial data with universities and NGOs who are conducting environmental or cultural resource research at their installations.

I&E Geospatial Data Assets Are Used and Shared for Diverse Purposes

Military I&E geospatial data assets are being used and shared for many diverse military purposes. Mission areas, such as emergency planning, response, and recovery; environmental management; homeland defense; military health; antiterrorism and force protection; morale, recreation, and welfare; public affairs and outreach; public safety and security; logistics; warfighting; and training all use I&E geospatial data assets in many different ways. Chapters Three and Four and the appendix give many different examples how I&E geospatial data assets enable these and other missions.

To better understand how the data assets are being shared for all these diverse purposes, Table 2.6, presents nine of the basic CIP GIS data layers/categories discussed above and identifies examples of how they are used.

The examples in this table provide different functional applications to illustrate the range of missions that use the same GIS dataset. An X in this table means the dataset would likely be used to help support the application if the dataset were available. Whether a dataset is

Table 2.6
Examples of How Datasets Are Used to Support Multiple Functional Applications

Application Example	Base Boundaries	Buildings	Roads	Runways and Taxiways	Ground Training Areas	High-Resolution Imagery	Utilities	Hydrology and Water Bodies	Vegetation
Site-planning at installation	X	X	X	X	X	X	X	X	X
Managing installation infrastructure	X	X	X		X	X	X		
Planning an air show	X	X	X	X		X	X		
Assessing the environmental impact of a new range	X	X	X	X	X	X		X	X
Managing bird species of concern	X	X	X	X	X	X		X	X
Explosive safety planning	X	X	X	X	X	X	X		X
Assessing installation force protection vulnerabilities	X	X	X	X	X	X	X	X	X
Preparing for hurricanes	X	X	X	X		X	X	X	X
Planning a military training exercise	X	X	X	X	X	X		X	X
Assessing the spread of Lyme disease	X	X	X		X	X		X	X
Meeting with the public about base noise concern	X	X	X	X	X	X		X	X
Assessing aircraft beddown for contingency operation	X	X	X	X		X			
Total	12	12	12	10	9	12	6	8	9

actually used would depend on the specifics of the application. For example, if the site-planning example was for a large facility, such asa new residential housing complex, then all the GIS dataset layers would probably be used to help analyze where to place it and what its effect might be. However, if the siting decision was for a small storage building near a training range in the middle of a large installation and not near any other structures, then the installation boundaries, other buildings, runways, and streams data layers probably would not be needed. Similarly, if assessing the environmental impact of building a new ground training range, ground training areas would obviously be important. If an air training range is being assessed, runways need to be considered. But if a ground training range not near a runway is being assessed, runways would probably need not be considered. If buildings are on or near the training range however, this geospatial dataset may be useful. Data on utilities would not likely be used unless they were being built for the range or were near bird species of concern, such as the bald eagle, which sometimes is killed by accidentally flying into power lines. The total number of Xs down each column shows that these basic GIS datasets can be useful for many different applications and for each application, many other types of datasets, especially mission-specific ones, would be used as well.

The scale of the application also affects the data needed. If an application is for a regional-level or national-level assessment, the user would usually not need such detailed data, preferring instead aggregate data for an installation or even a region. For example, when assessing the spread of Lyme disease across the United States, a user may just want to see dots representing installations that are sized and shaded based on the number of Lyme disease incidents at that installation rather than specific location information about where on the installation Lyme disease has been detected.

This table demonstrates another important point: Many basic I&E geospatial datasets are used for so many different purposes that it can be difficult to separate a "data use" example from a "data sharing" example. For example, the roads and boundaries of an installation are used in many different applications so the data on them are shared.

Therefore, throughout this monograph, we discuss both the use and sharing of I&E geospatial datasets, because in almost every application "use" involves at least one and usually multiple geospatial datasets that were shared in some way.

I&E Geospatial Data Assets Are Used and Shared in Many Ways

I&E geospatial data assets are physically used and shared many ways, from sophisticated web systems, to non-web network sharing, to burning CDs of data and hand-delivering them to someone.

Web-Based Sharing Systems

Military web-based systems used for sharing I&E geospatial data can be classified in three ways: by the range of data within the system and who has access; by the type of web viewer; and by the literature classification of spatial portals. Spatial portals are web-based gateways through which users can disseminate, discover, and access geospatial information from different sources.[8]

We discuss the first two here and the third in Chapter Six, because each provides a unique insight about these important data-sharing mechanisms and are so key to current and future sharing of I&E geospatial data assets.

The first way to classify military web systems is by the range of data within the system and who is allowed access. Some geospatial web systems are designed to view data for installations throughout the world for an entire Service or all of OSD, whereas others are limited to viewing data within a base, regional, or functional command.

DoD web systems are being developed so that users can access installation data across the entire United States or world. Sample national and worldwide web-based systems are shown in Table 2.7. Often, such systems contain only general data and have broad access. For example, the DISDI Viewer, as discussed above, allows users to view

---

[8]   Tang and Selwood (2005).

Table 2.7
Sample DoD National and Worldwide Geospatial Web-Based Systems

Portal Name	Short Description	Sponsor	Who Can Access It	Sample Data Assets
DISDI Portal	Installation data from the Service I&E geospatial data repositories	DUSD/I&E DISDI Office	Anyone with .mil access and DoD-issued credentials (Public Key Infrastructure (PKI) certificate; i.e., a military common access card (CAC)	IVT data for the United States, point base locations from DoD (2005a), and DoD Vector Smart Map (VMap) data for the world
U.S. Air Force GeoBase, Air Force Portal (Global Combat Support System (GCSS)–AF)	U.S. Air Force installation mapping and visualization	HAF/GIO	Anyone with an Air Force portal account and .mil access	Base boundaries, imagery, buildings, roads, and runways
Army GIS Repository (GISR)	Spatial data warehouse for storing installation geospatial information	ACSIM	Anyone with an Army Knowledge Online (AKO) account	1- and 5-meter imagery, base boundaries, buildings, roads, and ranges
USMC GEOFidelis Portal	Basic geospatial data for all Marine Corps installations	Headquarters Marine Corps Installations and Logistics Department	Anyone with .mil access and permission of the Service GIO	Imagery, base boundaries, roads, buildings, ranges, environmental hazards, cultural resources, and flora and fauna
U.S. Navy GeoReadiness Repository (GRR)	Foundation geospatial data for Navy installations	Naval Facilities Engineering Command, Base Development	Permission of the Navy Geospatial Information and Services program manager (GI&S PM) or a regional GI&S resource center	Imagery, base boundaries, roads, buildings, ranges, environmental hazards, cultural resources, and flora and fauna

Table 2.7—continued

Portal Name	Short Description	Sponsor	Who Can Access It	Sample Data Assets
Palanterra	U.S. geospatial information to support homeland defense and security	NGA	Unclassified version requires .mil access and need-to-know approval and classified versions require clearances	Jurisdictional boundaries, including base boundaries, buildings, roads, water features, topography, and high-resolution imagery

NOTE: The Regional Shore Installation Management System (RSIMS) is now part of this broader GRR system, but in the past there were different regional versions of RSIMS, such as RSIMS version 2. Because of this, throughout this monograph we discuss different regional RSMIS examples.

general data, such as installation point locations and boundaries, government jurisdictional boundaries, wetlands, and one- and two-meter imagery for U.S. military installations throughout the world. Anyone with .mil access and a Common Access Card (CAC) can view such geospatial data. However, this is not always the case. Some international and national systems contain more detailed installation data. The Navy has been developing the Regional Shore Installation Management System, an Internet map viewer that contains detailed and extensive Navy installation data. For example, many of the Navy EFD South installations have over 100 GIS layers in the system. In January 2006, this system became accessible Navy-wide, i.e., a global GIS system for all installations that can be used for a variety of functions including installation planning and base development. All installation GIS data are in this system. Other such systems have more restricted access. Palanterra is a national web-based system for homeland security and defense; its detailed geospatial data have both a more general access system that allows access by .mil users and nonmilitary emergency response personnel with a need to know. It also includes classified versions with even more restricted access.

The Services are also developing web systems that are limited to showing data about an installation, region, or functional command. Usually, users can access these systems only if their computers operate inside the installation, functional command, or regional Intranet system. Information security firewalls often restrict access to anyone outside this network. These systems also primarily provide geospatial information for that installation, region, or functional command. For example, Langley AFB, Virginia, has a system called "Langley Geo-Base MapViewer." This web site contains CIP data and other GIS layers for the installation. It is accessible to anyone inside the base firewall system, i.e., on an installation with NPRNET access (a .mil account). Off-base military contractors cannot access this system because of the installation firewall. However, such access restrictions do not always exist. For example, the USAREUR Integrated Training and Management (ITAM) program, "ITAM Mapper," is a web-based system that is used to view and assess training area geospatial

Table 2.8
Sample Service Installation, Regional, and Functional Web-Based I&E Geospatial Systems

Portal Name	Short Description	Who Can Access It	Sample Data Assets
		Installation examples	
Ramstein AB ARCIMS Viewer	Basic and advanced viewers; the advanced viewer has specialized tools for viewing and using I&E data	Available through the AF portal to anyone with portal access; the advanced viewer is limited to internal base use and password restrictions apply	CIP data, Environmental Planning and Assessment Tool, Asbestos Management System, and Emergency Response Tool Suite
NAS Patuxent River RSIP Viewer	I&E geospatial data assets for operating force support, community support and base support	Users with access to NAS Patuxent River Intranet	Interactive maps for air operations and facility support, public safety, and environmental management
Camp LeJeune IGIR	A centrally managed, enterprise GIS that integrates base GIS data into one integrated, shared resource that supports all mission-driven business processes	Different functional users across the installation	Buildings, transportation, flora and fauna, utilities, imagery, and military operations GIS datasets (see Table 2.1)
		Regional and functional examples	
IMA[a] Korea Region Office (KORO) GIS Repository	Basic viewer and special tools	Korea headquarters IMA staff and customers, and Army Camps throughout Korea	Web Evacuation Permit System (W-EPS), Web Siting tool (W-Site), and Web Planning Tool (called W-Plan)
USAF AMC MAJCOM GeoBase MapViewer	To view basic installation CIP data	Anyone with USAF AMC web access	CIP data
USAREUR ITAM Mapper	To view training area spatial data for USAREUR's training areas	Anyone with a .mil domain and access	Aerial imagery, digital topographic data, ranges and other training areas, land rehabilitation sites, roads, and nature protection areas

[a] Late in 2006, Installation Management Agency (IMA) was renamed Installation Management Command (IMCOM).

data for almost 100 of USAREUR's training areas. Anyone with a .mil site can access it. In the future, USAREUR ITAM will require an AKO logon.

Additional Service installation, regional, and functional web-based system examples are shown in Table 2.8. Second, web-based systems can be classified by the type of I&E geospatial data web-viewing service that they provide, from basic to more sophisticated geospatial viewers. Many organizations are developing basic installation geospatial data web viewers for non-GIS users. Such systems are usually accessible by many users either at the installation or throughout the Service. They often include standard and simple GIS functions, such as panning and zooming, identifying functionality, calculating distances, and buffering an area. For example, Langley AFB's Langley GeoBase MapViewer includes a basic installation geospatial data web viewer that is oriented toward non-GIS users throughout the installation.

Sophisticated web viewers are also being developed, often by the same organizations that have the basic viewers. These systems have more extensive data and functions, such as focused query and analysis tools and applications for specialized users. They usually have more restricted access, allowing access only to users with a need to know to. These systems may allow the downloading of special applications and often link directly into other installation functional databases. Langley AFB's MapViewer also has more sophisticated functionality for those with a need to know, such as a facility utilization tool. With this tool, those with a need to know can examine floor plan details and work order requests.

Many military organizations begin by developing a basic installation geospatial data web viewer and then, over time, they add extra functional and special applications as funding and interest dictate.

Nonmilitary web viewers also provide information about and possible direct access to I&E geospatial data assets. For example, the Federal Geographic Data Committee (FGDC)[9] has developed the Geospa-

---

[9]   The FGDC is an interagency committee that promotes the coordinated development, use, sharing, and dissemination of geospatial data on a national basis (www.fgdc.gov).

tial One Stop portal,[10] so that anyone can access geospatial information from federal agencies and a growing number of state, local, tribal, and private agencies through one comprehensive web portal. This portal also includes very limited information about U.S. military installations.

Non-Web-Based Methods for Sharing I&E Geospatial Data Assets

Geospatial data assets are also shared through desktop applications that use I&E geospatial data. Desktop systems can run faster and can more easily deal with large datasets on the desktop than can distributed systems, which access these datasets through the web; this is especially important when dealing with larger-scale datasets. Such desktop systems include stand-alone map viewers, GIS toolsets, and customized software for specific mission needs. For example, the USAREUR ITAM program developed the USAREUR "ITAM Viewer" for Army soldiers to use if they did not have easy access to their web-based ITAM Mapper. ITAM Viewer is a 2-DVD set used to view USAREUR training areas. The system is loaded onto the user's desktop. It provides access to aerial imagery, topographical maps, and vector data for more than 80 USAREUR training areas. Viewer features include zoom, measuring lines or areas, printing, and exporting to PowerPoint. This system is also easy to distribute at conferences. The USMC's GEOFidelis Tool Set, as discussed above, is another good example of a desktop toolset.

DoD organizations, especially installations, also share I&E geospatial data assets through non-web-based network systems. These networks are primarily used for sharing GIS datasets among GIS programmers who need the same GIS datasets for their mission applications. Such systems are aimed at GIS developers and they are focused on directly downloading GIS files. For example, at Fort Irwin, California, the ITAM and environmental staffs use the installation Intranet to share GIS datasets detailing vegetation and erosion information. FTP sites and email are other avenues for transferring GIS datasets and other I&E geospatial data assets, such as maps generated in PDF.

---

[10] See www.geodata.gov.

Geospatial data assets are also shared by means of CDs and DVDs, such as imagery and vector GIS datasets. Many geospatial data asset developers whom we interviewed said that they still used such "Sneakernet" methods for sharing, especially for sharing such large data file products as high-resolution imagery. Sneakernet is also used to work around firewall restrictions and to transfer unclassified I&E geospatial data into classified systems.

Geospatial data and information are also shared extensively in non-dataset forms such as official documents, maps, and PowerPoint briefings. Many DoD business processes that depend on geospatial data use these data to routinely support the documentation of these processes. For example, in the environmental area, GIS-generated maps are used in Integrated Natural Resources Management Plans (INRMPs), Cultural Resource Management Plans, and official NEPA process documents. Another common way that I&E geospatial data are shared is by means of documents from specialized mission studies that use GIS datasets in analysis. Such documents are often prepared by a contractor or even a university engaged in analyzing an issue for an installation.

I&E geospatial data and information are also shared in videos and simulators, often for training or other educational purposes. For example, GIS staff members at the environmental management office at USMC Camp Butler, Okinawa, Japan, have created a 3-D island fly-over prototype demonstration of the island and installation grounds as an environmental educational and public relations video.

Geospatial data assets are also collected and disseminated through field technology applications, such as hand-held GPSs or field computers that have geospatial applications on them. For example, to help assess the effects of building seven new training ranges because of BRAC 2005, Fort Benning natural resources staff members are conducting a forest inventory using hand-held GPS-enabled computers that have a forest inventory program on them. Detailed information in the field is being collected about the tree species, ages, size, etc. The field data are being transferred into the installation GIS system for analysis and use. At Dyess Air Force Base, Abilene, Texas, hand-held GPS units are being used to inventory trash and recycle dumpsters for contract maintenance purposes. Trimble GeoTX GPS units with customized menu GIS-based entry systems are used to enter data from the field into this hand-held system and then to transfer the data into the installation GIS system.

This chapter has defined what I&E geospatial data assets are, how installations are the main creators and maintainers of such assets (although other organizations within the Services and other parts of DoD also create and maintain some of these assets), and how the current users of these assets are spread across and even outside DoD. This chapter has also explained the many different ways that I&E geospatial data assets are shared and for what purposes. The next chapter describes how I&E geospatial data assets support many different missions.

CHAPTER THREE

How Do I&E Geospatial Data Assets Enable Diverse Missions?

During our research, we found that I&E geospatial data assets enable many diverse missions throughout DoD in many different ways. Any mission function that involves location information could use geospatial data to help track, manage, view, or analyze that information. This chapter discusses the main mission areas that are using I&E geospatial data assets within different parts of DoD.

For discussion purposes, we defined 16 mission area categories based on traditional installation and warfighting operations that use geospatial data assets. Main mission areas where I&E geospatial data assets enable traditional installation mission uses, mainly business-related mission uses, include

- base planning, management, and operations
- emergency planning, response, and recovery
- environmental management
- homeland defense, homeland security, and Critical Infrastructure Protection (CIP)
- military health
- morale, recreation, and welfare: enhancing quality of life
- production of installation maps
- public affairs/outreach
- safety and security
- strategic basing

- training and education
- transportation.

Main mission areas where I&E geospatial data assets enable traditional warfighting operations include

- command, control, communications, and computer (C4) systems
- logistics
- warfighting strategic planning, policy, and assessments
- warfighting operations

It is important to distinguish business missions from warfighting missions. Therefore, this chapter focuses on the traditional installation and business missions and the next addresses warfighting missions.

These 16 categories are not mutually exclusive; they overlap. The categories were chosen to highlight categories where geospatial information is most applicable.

I&E geospatial data assets are used and shared within some mission categories more than in others. The three largest mission application areas were environmental management; base planning, management and operations; and training, which makes sense since these were the first to develop and use geospatial data. We easily identified over 100 environmental application examples and over 70 in each of the other two categories. These three mission areas also are large; others are more focused. For example, strategic basing is a narrow category, referring to such strategic basing issues as BRAC and other activities to realign or close U.S. installations around the world.

To illustrate how the I&E geospatial data assets are shared to support these many diverse mission areas, Table 3.1 shows how four installations share I&E geospatial assets to support the 16 mission areas: U.S. Army's Aberdeen Proving Ground, Maryland; USMC Camp Butler, Okinawa, Japan; U.S. Navy's NAS Patuxent River, Virginia; and USAF Ramstein AB, Germany.

An X in this table means that we know the mission area has used installation I&E geospatial data. These uses represent only a minimal set for the installation, since the information is based on a limited number

Table 3.1
Samples of Which Mission Areas at Four Installations Are Supported by I&E Geospatial Data Assets

Mission Area	Aberdeen Proving Ground	Camp Butler	NAS Patuxent River	Ramstein AB
Installation-related uses				
Base planning, management and operations	X	X	X	X
Emergency planning, response, and recovery	X	X	X	X
Environmental management	X	X	X	X
Homeland defense, homeland security, and Critical Infrastructure Protection (CIP)	X	X	X	X
Military health	X			X
Morale, recreation, and welfare: enhancing quality of life	X	X	X	X
Production of installation maps	X	X	X	X
Public affairs/outreach	X	X	X	X
Safety and security	X	X	X	X
Strategic basing	X	X	X	
Training and education	X	X	X	
Transportation	X	X	X	X
Warfighting operational uses				
C4 systems	X			X
Logistics				X
Warfighting strategic planning, policy, and assessments		X		X
Warfighting operations	X		X	X
Total number of these 16 missions known to use I&E assets	14	12	12	14

of staff interviews. Only two to six people were interviewed at each installation, so not all the potential mission users were contacted. At some installations, the data are likely being used for missions not identified in the table.

This table shows only a minimum set of uses of I&E geospatial data for only 12 to 14 of the 16 mission areas at all four installations. Clearly, these data assets support most of the mission areas at all four installations, including important warfighting missions. Looking across all the installations, every mission area has been supported by installation I&E geospatial data. The four installations in the table were chosen because they use and share I&E geospatial data assets extensively and many other installations do not yet have such widespread use and sharing.

To illustrate how the I&E geospatial data assets enable different missions within different parts of DoD, we next present diverse examples organized by organization type:

- installation-level uses
- applications by regional and functional organizations and commands
- service headquarters applications
- OSD applications
- uses in other parts of DoD
- uses by organizations outside DoD.

For each organization type, two to four examples are provided to show how different organizations and people use I&E geospatial data assets to support different missions. The appendix provides more examples by each of the 12 mission areas to more fully illustrate the diversity of the applications. In fact, it provides over 130 application examples.

Installation Level

Installations use I&E geospatial data assets for installation missions in diverse ways because many have a long history of using geospatial data within certain mission areas. To illustrate, we present four types of applications from an installation for each Service: Fort Hood training simulator; Camp Butler storm water management and fire recovery; NAS Patuxent River base services management; and Langley AFB's floodmap management tool for emergency response.

At the installation level, I&E geospatial data assets have been used to develop, manage, and operate training ranges. Imagery and other GIS data have even been integrated into training simulators, which can reduce the amount of time that soldiers need to train on a range. This is an important mission effect, since there are many pressures on training ranges, such as from encroachment, and the need for training range space is expected to increase in the future.[1] For example, at Fort Hood, Texas, training range GIS aerial and topographic data are used in tank and aviation simulators, which helps orient the soldier and saves valuable time on the training range. This practice has cut by about one-third the amount of time that helicopter pilots need to spend on the gunnery range. A-64 Apache helicopter pilots fly the Fort Hood simulation model before going out to fly and shoot on the gunnery training range. Previously, they would make an initial flight pass first at the gunnery range and then fly and shoot, but with the realistic installation simulator, they no longer need to make that first pass.

Environmental management staff members at USMC Camp Butler, Okinawa, Japan, have created a detailed 3-D geospatial model of the drainage on and around Camp Butler to better manage storm water runoff. The GIS data includes aerial photography, building locations, streets, elevation data, and even the manhole cover locations. See Figure 3.1 for an image from the 3-D storm water runoff model of Camp Butler.

This 3-D modeling system was used to help address water runoff problems after an accidental range fire in early April 2005 in the Central Training Area (CTA) at Camp Butler. The CTA has fine red clay soil and erosion can be detrimental to nearby coral reefs. This fire thus created a large amount of political attention for the USMC because of erosion concerns. Using GIS-based watershed modeling, the Camp Butler environmental GIS staff mapped the burn area and analyzed the potential erosion problem. A U.S. Department of Agriculture (USDA)

---

[1] For example, today a Stryker brigade combat team has a doctrinal battlefield footprint of 40 x 40 kilometers (1,600 sq km) and the future force is expected, by one estimate, to have a 75-kilometer radius (17,671 sq km) doctrinal footprint requirement (Knott and Natoli, 2004, p. 12).

Figure 3.1
Picture from the Camp Butler 3-D Storm Water Runoff Model

SOURCE: Image courtesy of Camp Butler environmental management staff.
RAND *MG552-3.1*

soil scientist, stationed at Camp Butler, suggested aerial hydro-seeding of the burn area to promote rapid vegetation regrowth, which would reduce the runoff. (Because the CTA is a live fire range, seeding could not be done from the ground.) Using this analysis, GIS staff calculated slopes greater than 30 percent and 20 percent to determine which areas to seed. Burn areas were hydro-seeded by helicopter using a seed and fertilizer mixture. 3-D GIS modeling was also used to identify potential runoff locations so that countermeasures could be taken on the ground along the road, with the assistance of the Japanese government.

Because of geospatial analysis, erosion was reduced and relations with the Japanese government were improved. The analysis also helped reduce fire damage claims against the U.S. government, as will be discussed more in Chapter Seven.

At NAS Patuxent River, GIS analysis is used to more efficiently and fairly allocate janitorial and utility costs to tenants and to help oversee janitorial contracts. Within the GIS, janitorial code numbers are overlaid on building floor plans so the user can see who is supposed to clean which office areas. Different offices in a building are owned by different tenants, such as operations and maintenance (O&M) and research, development, test, and evaluation (RDT&E). Installation geospatial staff members calculate which offices are cleaned and divide hallways/stairways evenly among the tenants. Then they calculate the space in the floor plan, determining the net square footage for each customer. Those who use only one-quarter or one-third of a building want to pay for only one-quarter or one-third of the utilities and cleaning services, which this system ensures they do. Besides allocating janitorial and utility costs more fairly, this system also has improved management staff's communications with tenants. Tenants more readily understand and accept their bills with this space allocation method.

Langley AFB has developed a web-based tool to better plan and respond to floods. The base is on low-lying ground at the mouth of the James River near the entrance to the Chesapeake Bay. Flooding from Atlantic storms is a recurrent problem at the installation, and it is critical to anticipate the effect of flooding from forecasted storms. The Langley GeoIntegration Office has created a web GIS-based FloodMap Tool to support emergency response planning and real-time response. This tool was developed after significant flooding occurred after Hurricane Isabel hit in September 2003. See Figure 3.2 for a sample scenario within the tool. Each building's color shows the expected flooding effect on that building.

Commanders, emergency responders, base managers, and other staff across the base used this tool in September 2005 during Hurricane Ophelia. As Ophelia moved closer to the installation

> building managers moved to secure their buildings, and they were able to get real-time information on the event and how it affected their buildings from the FloodMap Tool. The FloodMap Tool interface allowed these personnel to view a map of Langley AFB and their area of concern or search for a specific building number

Figure 3.2
Sample Flood Scenario Within the Langley AFB FloodMap Tool

SOURCE: Image courtesy of the Geospatial Information Office, Langley AFB, 2006.

RAND MG552-3.2

and then magnify the view to see a map image of their building and the surrounding area in relation to the flood waters. They could also print out a report of affected buildings with a map image. In record time, sand bags appeared in front of vulnerable building doors, equipment and personnel were prepared for evacuation, and base managers and commanders had a common information picture of the entire installation.[2]

---

[2]   McSherry and Hardy (2006, p. 3).

At the height of the incoming storm, within a few hours of a base-wide email announcement giving the Internet address for the Flood-Map Tool, nearly 3,500 base personnel had accessed the service online. Having this tool online also helped reduce map requests at the Langley GeoIntegration Office.

Applications by Regional and Functional Organizations/ Commands

Different regional organizations, whether the Navy's Northwest region of the United States, the U.S. Air Force in the Pacific (PACAF), or U.S. Army ITAM Europe, use I&E geospatial data assets to help in regional planning, analysis, and operations for multiple installations. For example, the Navy EFD South used installation GIS data to create "damage profiles" for base structures after Hurricane Katrina. People deployed in the field entered damage information into the GIS onsite, including information about damage to buildings, runways, and roads from flooding and winds. These damage profiles were sent to Naval Facilities Engineering Command (NAVFAC) headquarters and the Commander Navy Installations (CNI) and were used to help with regional reconstruction planning, demolition, and the siting of new facilities. Damage profiles were created for numerous installations including NAS Pensacola, Florida; Construction Battalion Center Gulfport, Mississippi; NAS New Orleans, Louisiana; NAS Meridian, Mississippi; and Stennis Space Center, Mississippi (which is shared by the Navy and the National Aeronautics and Space Administration (NASA)). Such activities to help assess and restore installation buildings, facility assets, and infrastructure are common emergency recovery uses of I&E geospatial data assets at an installation level, for a region, or a functional command. Similarly, in the USAREUR ITAM program, "ITAM Mapper" and "ITAM Viewer" are used by trainers and soldiers to view and assess training area geospatial data for almost 100 of USAREUR's training areas. Commanders use these systems to help plan training and soldiers use them for training orientation, for land navigation, and for viewing an area before entering it. See Figure 3.3 for a sample map view of the trend analysis tool in the ITAM Mapper.

Similarly, functional commands and organizations, such as the U.S. Air Force Space Command, U.S. Air Force Air Combat Command, and USA-CHPPM, use I&E geospatial data assets to help in their functional missions. For example, the USAF Air Mobility Command has used installation I&E geospatial data for utility and housing privatization planning and decisionmaking processes across multiple bases. Similarly, in Korea, the U.S. Army's 18th Medical Command is using GIS to examine regional disease vector trends and preventive measures that could be taken. Army health analysts at 18th MEDCOM caught rats, mice, and mosquitoes to sample the population dynamics, as

Figure 3.3
Sample View of a Map Within the USAREUR ITAM Mapper

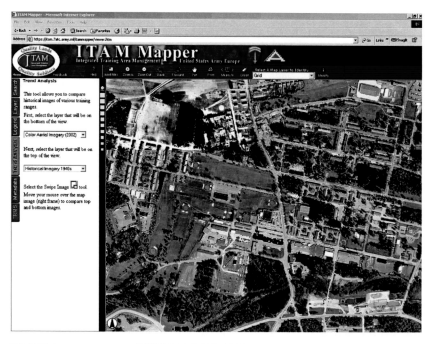

SOURCE: Image courtesy of USAREUR ITAM, 2006.
RAND *MG552-3.3*

well as disease rates, and have placed locations of malaria, insect bites, rats, mice, mosquitoes, and other disease vector information in a GIS. They use this information to determine preventive health measures, for example, if a training area has a particular infestation, to make sure precautions are taken there.

Service Headquarters Application Examples

At the Service headquarter, the headquarters GIOs, as well as other headquarters organizations, also use I&E geospatial data assets. However, in many cases, these are recent applications, mainly because the I&E data assets and capabilities became available at the headquarters level only recently. In fact, the availability of IVT data has facilitated more headquarters applications, as will be discussed in Chapter Five.

Many headquarters GIO I&E geospatial data applications support requests from senior Service management as well as OSD requests. An example of a headquarters GIO response was the use of I&E geospatial data assets from the USMC to provide data to the Navy Treaty Implementation Program to help with nuclear treaty verification and inspections.

USAF headquarters air and space operations branch uses I&E geospatial data assets to respond to congressional inquiries, to monitor hurricane response preparations at headquarters, and even to address noise complaints. For example, the USAF used to receive two to three noise complaints per month from people in national parks. Someone would complain at the park, the complaint would go up the chain of command to the National Park Service (NPS) headquarters in Washington, D.C., and then headquarters USAF would be contacted. USAF headquarters operations branch and the NPS developed a "United States Air Force and National Park Service Western Pacific Regional Sourcebook," a communication guide so that the local NPS manager could call the local USAF base manager about the noise complaint. The guide includes detailed maps and contact information showing the locations and phone numbers for each base and national park. See

Figure 3.4 for a sample map page from this book for Channel Islands National Park and Channel Islands Air National Guard in California.

This approach has solved problems at the local level, avoiding headquarters staff work. Since this book was produced, no noise complaints from national parks have been received at Air Force headquarters.

Figure 3.4
Sample Map Showing Channel Islands Air National Guard Station and Channel Islands National Park

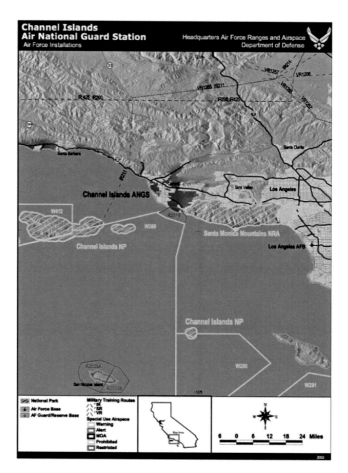

SOURCE: National Park Service and U.S. Air Force (2002).
RAND MG552-3.4

Office of the Secretary of Defense Application Examples

Other parts of OSD, such as OSD Health Affairs, OSD Homeland Defense, and the Deputy Assistant to the Secretary of Defense (DASD) Chemical Demilitarization and Threat Reduction (CDTR) staff members, have also used I&E geospatial data assets to help in their analysis, planning, management, and operations. Many of these applications are more recent and are taking advantage of IVT data, as is discussed in Chapter Five. We present two OSD application examples here from the environmental and explosive safety domains.

The OSD Office of the Deputy Under Secretary of Defense for Installations and Environment has been working with Florida, Georgia, South Carolina, North Carolina, and Alabama to form the Southeast Regional Partnership for Planning and Sustainability (SERPPAS). SERPPAS is a pilot effort to develop a working regional partnership between DoD, the Southeast states, and other stakeholders. Their agreed-upon mission is "To seize opportunities and solve problems in value-adding ways that provide mutual and multiple benefits to the partners, and sustain the mission and secure the future for all the partners, the region, and the nation."[3] The DISDI Office provided geospatial data using IVT data and test and training range data from other sources. DISDI data defined the DoD installation footprint within this partnership area. The Conservation Fund, a conservation NGO, combined these installation data with ecological and land-use datasets to produce maps and help assess encroachment and potential buffer areas around installations and to help develop conservation corridors of mutual benefit. These maps have been instrumental in helping identify potential areas to focus on within the region; see Figure 3.5 for a sample map.

It is important to note that DISDI has been a key facilitator in the SERPPAS collaboration process. DISDI staff members have provided technical assistance, including products; briefed audiences on the role of geospatial information in the process; made sure the partnership uses authoritative DoD data; and worked with North Carolina's Geo-

---

[3] SERPPAS Meeting Summary January 11–12, 2006.

Figure 3.5
Sample Map Used in the SERPPAS Process

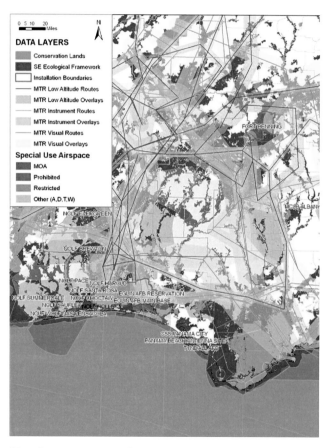

SOURCE: Map courtesy of the DISDI Office, 2006.
RAND *MG552-3.5*

spatial Information Council to coordinate data sharing. DISDI staff members have also worked with other Southeast states to help them improve the communication between state geospatial coordination bodies and DoD agencies.

Next, we present a DoD explosive safety application example. Every U.S. military installation, including permanent and contingency bases, is required to develop "Explosive Safety Site Plans" (for each of their explosives storage and operating facilities), which describe and show how the installation meets DoD explosive safety standards to minimize the risk of explosive accidents. The DoD Explosives Safety Board (DDESB) in Washington, D.C., reviews the installation site safety plans submitted to ensure that they comply with DoD standards. DDESB staff members use recent imagery of an installation and installation boundaries to help review site plans and identify risks or violations.

The DDESB also conducts surveys at installations to validate that site plans were enacted as stated. The validation process includes assessing whether a specified facility was constructed, placement of storage facilities, and the amount of explosives storage within a given facility. Before traveling to specific installations for surveys, the DDESB is generally given a hard copy map of the installation and occasionally a CD containing GIS data. DDESB staff members would like to use more digital I&E geospatial data assets, such as the DISDI Portal, to help in these surveys. Such assets help in several ways. They help DDESB prepare and become familiar with the base and to make travel arrangements, since they can quickly see where the bases are in relationship to each other and major cities. Since DDESB cannot visit every installation, the assets also can help DDESB staff members prioritize their visits by helping them to identify potential problem installations, such as ones with new encroachment problems near explosive storage areas. The use of such assets can save them time and money and can help them find previously unidentified violations of explosive safety standards that put the mission at risk.

Uses by Other Parts of DoD

Other parts of the Department of Defense, such as NGA and the USACE research and development (R&D) labs, also use I&E geospatial data assets. NGA is a main DoD user of I&E geospatial data assets in such areas as producing training maps for the warfighters and homeland security and defense applications. Project Homeland is a collaborative effort to provide geospatial information to federal, state, and local government agencies for homeland planning, mitigation, and response so that the U.S. government can more effectively respond to emergencies, whether a terrorist attack or a natural disaster. NGA has the federal lead on this effort with USGS support. A Project Homeland Colorado pilot was started in the Colorado Springs area and has been extended to the state of Colorado. In fact, Colorado is using the pilot for its Emergency Operations Center and the Palanterra system design for its web-mapping solution. The pilot project is developing data-sharing models and agreements to facilitate effective sharing across federal, state, and local government agencies. Several military installations are active in this pilot, including Peterson AFB, the USAF Academy, and Fort Carson. DISDI is helping to develop the military data-sharing agreements and data model for this pilot.

An R&D example developed by the USACE Construction Engineering Research Laboratory (CERL) is the U.S. Army's Fort Future. Fort Future is a framework for modeling and simulation for installation planning. It includes a suite of capabilities that integrates different tools, including a range tool, facility composer for designing buildings, and the Land-Use Evolution and Impact Assessment Model (LEAM), for assessing potential landscape changes near an installation. These tools can be used by diverse installation and other Army staff to help with assessments at different levels of scale from the building level to regional planning.

How Organizations Outside DoD Use I&E Geospatial Data Assets

Other federal agencies, state and local governments, universities, NGOs, and even industry use I&E geospatial data assets in such areas as environmental management, emergency response, and homeland defense.

More and more state agencies are using military I&E geospatial assets to help with geospatial-based homeland security and emergency response systems. For example, Pennsylvania is creating a homeland security/public safety geospatial web-based portal system for state agencies and data sharing with local governments and first responders. It is developing a secure geospatial portal for sharing, mapping, and analyzing information. This system will provide a homeland security common operating view that leverages a $15 million military investment in the Intelligent Road and Rail Information Server (IRRIS) technology. Capabilities will include a single application framework to visualize all threats and emergency incidents and will therefore play a critical role in the development of the Homeland Security Data Fusion/Intelligence Center for the state. Pennsylvania has acquired and is using military installation geospatial information in this system.

Nongovernmental and nonprofit environmental organizations and universities often use geospatial data to help the U.S. military address environmental issues, such as T&ES management and encroachment issues. For example, to limit the effects of urban and suburban sprawl on military training, the Army has developed a plan to "buffer" training lands from this encroachment and to maintain its ability to use and access the current training ranges within installation boundaries. The Army Compatible Use Buffer (ACUB) program allows federal funds to be used to form partnership agreements with county, state, or municipal governments as well as nonprofit organizations so that the partner can buy tracts of land or easements on lands that surround the installations. Conservation NGOs, such as the Conservation Fund and the Trust for Public Land, help develop such agreements, often identifying tracts of land near installations that would make suitable buffers. I&E GIS data have been used by such organizations to help develop ACUB priorities and documentation for several installations in the Southeast. These assets were useful in ACUB project developments at Camp Ripley, Minnesota; Camp Blanding, Florida; Fort AP Hill, Virginia; Fort Bragg, North Carolina; Fort Carson, Colorado; Fort Sill, Oklahoma; and Fort Stewart, Georgia. For a sample of a GIS map used in the ACUB process, see Figure 3.6, which shows an ACUB map for Fort Sill, Oklahoma.[4]

---

[4] For more examples of such maps, see U.S. Army Environmental Center (2005a).

Figure 3.6
ACUB Map for Fort Sill, Oklahoma

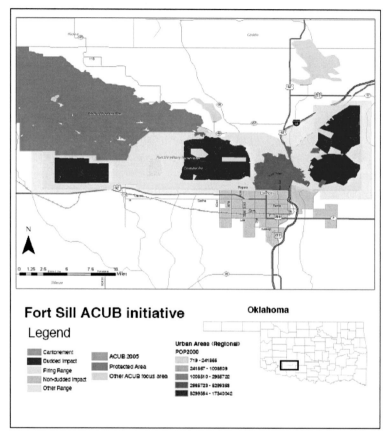

SOURCE: U.S. Army Environmental Center (2005a, p.17).
RAND *MG552-3.6*

CHAPTER FOUR

How I&E Geospatial Data Assets Enable Traditional Warfighting Operations

As mentioned in Chapter Three, I&E geospatial data assets also support warfighting missions across the Department of Defense. To help the discussion, the warfighting categories are grouped as in DoD Joint Staff Directorates, i.e., J1–J8, since this is how DoD is organized for warfighting. Our discussion focuses on mission application areas where geospatial information and expertise are most relevant. Three categories cut across all mission areas and they are not separated out in this discussion: J1 Manpower and Personnel, J2 Intelligence, and J7 Operational Plans and Interoperability. The categories discussed below are

- Command, Control, Communications, and Computer (C4) Systems (J6)
- Logistics (J4)
- Warfighting Operations (J3)
- Warfighting Strategic Planning, Policy, and Assessments (J5, J8).

Command, Control, Communications, and Computers (C4) Systems

I&E geospatial data assets support C4 systems[1] that are critical to warfighting operations by providing a common baseline that can be used

---

[1] C4 systems is a Joint Staff Directorate (J6) that advises the chairman on all C4 matters, guides the direction of the C4 community, and oversees support for the National Military Command System.

for planning, deploying, and operating various types of command, control, communications, computers, and intelligence (C4I) systems. The datasets can support long-term planning of physical site operations. In deployment, they support such activities as the proper routing of critical landline elements or the situating of communications antennas. In operating C4I systems, they can support terrestrial wireless communications systems by creating supporting datasets for a variety of specialized tools to assess localized communications dead zones (areas of low apparent signal strength) or the effect of possible interference sources on C4I activities.

At many U.S. military installations throughout the world, operational communications networks are being planned, deployed, and operated to support ongoing operational needs. I&E geospatial data assets are used to help support the laying out of such networks and the siting of physical elements of C4I systems such as antennas, landline elements, or special facilities. Often, the C4I staff use such information to determine the general locations for operational networks and then the installation geospatial staff use the information to locate and manage such networks. For example, the Navy EFD South GIS staff used installation GIS data to help site communications towers and satellite dishes at NAS Key West, Florida, for the C4I community. C4I staff also use I&E geospatial data on the internal layouts of buildings, such as at RAF Lakenheath, the largest U.S. Air Force–operated base in England. At this USAFE installation, the C4I staff used the installation GeoBase web viewer to examine the floor plans of different buildings. They also requested and received a CD of the floor plans to use in their own geospatial analyses.

I&E geospatial data assets are also useful in the development, testing, and application of battlefield command and control systems that use geospatial information. For example, at APG, GIS data helped support the R&D testing of unmanned aerial vehicles (UAVs) and the integration of their video feedback into battlefield command and control systems. These systems used installation digital terrain data. In addition, in certain circumstances, installation I&E geospatial data and knowledge about such data could help contribute to warfighting command and control situational awareness. For example, if Seoul,

South Korea, was attacked, I&E geospatial data and systems at nearby U.S. bases could be useful, especially if they had been engaging in extensive geospatial data sharing with the South Korean government.

Logistics

I&E geospatial data assets also are used to support military logistics. Logistics includes logistics support for strategic and contingency planning and operations. Often, systems that are useful for peacetime and installation logistics are also useful for warfighting.

The Army Transportation Command (TRANSCOM) Military Surface Deployment and Distribution Command (SDDC) Transportation Engineering Agency (SDDCTEA) develops and uses I&E geospatial data in geospatially based tools to help track military supplies and materials all over the world. For instance, SDDCTEA developed IRRIS, a secure web-accessible GIS system to monitor transportation logistics data and real-time tracking information. IRRIS currently has over 140 data layers and is accessible to 400–500 military users. Decisionmakers can use IRRIS to manage and track U.S. military freight and equipment in real time. Users can display layers that include aerial photos, rail lines, waterways, terminals, etc. The system also includes detailed maps of military installations, depots, and airfields, so that drivers know where to go when they reach their destinations. The geographic (locational) nature of the logistics information facilitates the use of GIS mapping by allowing users to visualize assets and to perform analysis (e.g., plume, buffer, route adherence, automated alerts, and notifications) and spatial queries. IRRIS has allowed SDDC to streamline military logistics, reducing the time and costs for military training and operations. IRRIS provides the military and DoD with greatly improved logistics situational awareness worldwide. It is especially effective for monitoring sensitive shipments. IRRIS incorporates real-time weather and transportation data from a private company. The U.S. Navy also has its own version of IRRIS for tracking ships from port to port.

Similarly, individual installations and regional commands use I&E geospatial data to help support warfighting logistics. For example, Ramstein AB installation geospatial data were used to support logistical operations in recent contingency operations, since this base is a key logistical hub for U.S. Central Command (CENTCOM). USAFE headquarters has also used I&E geospatial data assets within its classified GeoReach system (the USAF warfighting geospatial system) to help with logistics planning in contingency operations, since some key European permanent bases are a stopping point on the way to and from contingency operations, such as operations in Iraq and Afghanistan.

Warfighting Operations

Permanent military installation I&E geospatial data assets also help support current warfighting operations and plans (J3). I&E geospatial data assets are used to help activities related to the analysis and conduct of warfighting mobilization and deployments, base camps and other forward operating sites, combat operations, and post-conflict stability and reconstruction. Permanent installation geospatial expertise and applications help improve the safety, efficiency, and effectiveness of such operations. For discussion purposes, examples of supporting warfighting operations are grouped into four categories:

- combat and post-conflict operations
- force projection: supporting rapid deployment
- rapid basing and forward basing: tools and techniques from permanent bases
- specialized training and weapons testing for current operations.

Combat and Post-Conflict Operations

I&E geospatial data assets are useful for supporting combat and post-conflict operations, especially when the U.S. military has responsibility as an occupying power or as the leader of a coalition and needs to build up a database useful for meeting the needs of the local commanders who will be responsible for both security and maintenance of the civilian population. In addition, the data can be used to support longer-term reconstruction activities in the theater by providing a sound starting point for those activities.

Army and USMC warfighting and intelligence staff interviewed for our study stated that installation geospatial data on the urban environment, such as buildings, have more applicability for urban combat than in the past. One Service officer whom we interviewed discussed how urban warfare has become more important and its nature has changed. He discussed how during World War II, Berlin was reduced to rubble, but such extensive collateral damage is no longer acceptable and light infantry data requirements are different in the urban environment. Installation geospatial data analysts have the knowledge, expertise, standards, and applications for managing, using, and assessing detailed urban infrastructure data. All of these are now more relevant to the warfighter in combat.

For example, Joint Engineer Planning and Execution System (JEPES) is a theater-level warfighting engineering planning tool that collects and analyzes engineering information. Developers want to include applicable I&E data assets, such as useful I&E GIS tools, so that operational planners can use them in JEPES.

In 2005, to improve counterinsurgency activities in Iraq, the U.S. Army started using GIS to plot the locations of improvised explosive devices (IEDs) in urban areas as a prelude to conducting detailed geospatial analysis of the patterns and trends of their placement and to assess where triggermen can set up and where the United States could set up snipers to counteract possible insurgency attacks. I&E geospatial data assets, such as GIS tools used for calculating line-of-sight analysis in training exercises, could be used to help in such analyses, saving time and money.

Stability and reconstruction activities also could benefit from I&E geospatial data asset expertise, especially if such assets were used to help design, build, operate, and manage infrastructure. For example, in Iraq and Afghanistan, the U.S. Army and USMC help to rebuild roads, water, sewage, electrical, and trash systems. I&E geospatial expertise and tools for managing and building such infrastructure could help in such reconstruction activities. Expertise about such tools could also be taught to local people so that they can use them in reconstruction activities, thus strengthening long-term stability.

USACE geospatial expertise and knowledge from U.S. watershed analyses have been used to help develop a reservoir system simulation model for use in both day-to-day operational decisions and long-term water resource management studies in Iraq. The USACE is using its experience from modeling in U.S. watersheds to develop a model that will help restore key water flows throughout the country including the Mesopotamian Marshlands. Using this same approach in Afghanistan, the USACE and Afghanistan Engineer District have teamed up to develop a reservoir simulation model of the Kajaki Reservoir and other projects in nearby valleys.[2]

Force Projection: Supporting Rapid Deployment

I&E geospatial data assets are also used to more quickly mobilize and deploy U.S. troops to contingency activities and humanitarian assistance missions around the world. I&E geospatial data assets at strategically located permanent bases are especially useful for rapid deployment planning, operations, and logistics. Helping with forces deployment beddown decisions, i.e., where to place aircraft, troops, and equipment during the deployment process, is a typical application. For example, a four-star USAFE general used some of the Fairford AFB GIS data to help with aircraft beddown decisionmaking during Operation Iraqi Freedom (OIF). Installation geospatial information is also used to help develop and implement mobilization plans, as the Army did at Fort Bragg, North Carolina.

Rapid Basing and Forward Basing: Tools and Techniques from Permanent Bases

I&E geospatial data assets, especially expertise and tools to help manage and build training ranges and installation infrastructure, are also used to help plan, develop, manage, and operate U.S. base camps,

---

[2] For more details on these modeling efforts, see Gould (2004) and Gould and Hanbali (2004).

forward operating sites, and other field locations during contingency operations and other operational missions, both in host nations and in countries of conflict. Both the U.S. Army and USMC used geospatial data assets for these purposes. For example, at Camp Falugah in Iraq the Camp Commandant requested installation GIS infrastructure support. The warfighters did not have the appropriate geospatial skills and knowledge to use GIS information in setting up and running the camp. Temporary bases in warfighting situations benefit from sharing installation geospatial skills and applications. Installation GIS support is needed for facilities in contingency operations and humanitarian assistance missions.

Training ranges also are developed and used in contingency operations and benefit from I&E geospatial data assets. For example, in Iraq, the Army was trying to design new training ranges and wanted help from the range developers in the continental United States (CONUS). Army training range planning staff went to Iraq with the RMTK, a GIS-based toolset for analyzing and developing ranges, on a laptop and used it to help design the ranges and minimize the effect on local communities.

The USAF has also used I&E geospatial tools to help with contingency operations. For example, the USAF ACC Contingency Aircraft Parking Planner (CAPP) is a GIS-based system used to assess aircraft parking options at forward operating locations and other installations supporting ACC operations. The system determines the best aircraft apron size based on the number of aircraft and parking configurations. It uses the Spatial Data Standards for Facilities, Infrastructure, and Environment (SDSFIE), standards used by the I&E geospatial data community, so that data can be shared and used for other functional purposes. Besides supporting U.S. and overseas permanent installation operations, this geospatial data application has supported warfighting operations, such as at installations supporting operations in Afghanistan and Iraq.

Another key area where I&E geospatial data models and techniques are used at forward bases and sites is for force protection, critical infrastructure, and other safety concerns. Sharing geospatial expertise helps save money and time and improves safety and planning, thus helping save lives. For example, the ASHS program, a GIS-based application software tool to assess capacities for explosive hazards reduction,[3] has been used to help plan and manage explosives safety at deployed bases, such as installations in Saudi Arabia, and by CENTCOM, U.S. Central Command Air Forces (CENTAF), and PACAF to support operations in Afghanistan and Iraq.

Specialized Training and Weapons Testing for Current Operations

I&E geospatial data assets have also been use to conduct specialized training and testing of equipment and soldiers in direct preparation for current warfighting operations. At Fort Hood, Texas, installation GIS data were used for the "mission rehearsal exercise" to help units prepare for upcoming combat. This training exercise provides operational deployment training right before troops deploy overseas. At NAS Patuxent River, GIS staff members have supported the testing of warfighting equipment for current operations. For example, they supported weapon separation testing for bunker bombs for operations in Iraq.

Warfighting Strategic Planning, Policy, and Assessments

I&E geospatial data assets have also been used to help with warfighting strategic planning, policy, and assessments. This category combines the Joint Staff Directorate J5 Strategic Plans and Policies and J8 Force Structure, Resources, and Assessment functions. These areas focus on current and future military strategy, planning guidance, policy and war gaming, and other assessments to support warfighting planning. For example, J8 analysts use computer-aided models, war games, and politico-military seminars to assess threats to U.S. national security.

In Korea, for defense planning and wargaming, Army warfighters, analysts, and planners use installation geospatial I&E data to track seaport, railhead, airport, and runway information. Having this geospatial information about an installation's real property assets is impor-

---

[3] For more information on ASHS, see the explosive safety discussion in the appendix.

tant. Analysts and planners need to look at capacities and where to bed down troops supplies, etc. If part of a port or other transportation node is degraded, troops would need to be rerouted. Similarly, analysts at Camp Butler used I&E geospatial data assets to help with defense planning for a chemical or biological attack.

I&E geospatial data assets are also used in Europe within strategic and tactical wargames, simulations, and other warfighting planning and assessments. USAREUR uses installation terrain and other ITAM data in its simulations and the USAREUR WPC also uses such data in its wargames.

Analysts in the intelligence community, such as at the Defense Intelligence Agency, also use I&E geospatial data assets to assess threats to U.S. national security. An example of such uses is the DGINet (Distributed Geospatial Intelligence Network) tool/capability. DGINet is a system currently deployed throughout the intelligence community for web mapping, geospatial analysis, and information sharing of critical intelligence data. This information helps support planning for contingency operations and the Global War on Terrorism as well as other intelligence purposes. This distributed approach allows users to access data via web, open client, or GIS interface and dynamically discover the most recent data from a host of sources. DGINet dynamically provides geospatial visualization and analysis tools.

In summary, I&E geospatial data assets help support many warfighting missions, including C4 systems; logistics; strategic planning, policy, and assessments; and warfighting operations, such as facilitating more rapid deployment, better managing forward basing, and improving stability operations. Chapter Five discusses an IVT data-sharing case study, and Chapter Six, which discusses future use and sharing of I&E geospatial data assets, briefly discusses how and why the warfighting and intelligence uses of I&E geospatial data assets that were discussed in this chapter are likely to increase in the future.

IVT Case Study of Cross-Departmental Data Sharing

IVT was developed for the 2005 BRAC process by the IVT program office (which became the DISDI Office part way through the process). The IVT was designed only for "situational awareness" in the BRAC process, yet the data created for the IVT were used for much more.

IVT included detailed consistent GIS data for all four Services. The development of the IVT data was the first attempt to systematically produce consistent I&E geospatial data for OSD to use in high-level decisionmaking. When examining the uses and sharing of these data, we found that the IVT data benefited not just the BRAC process but also other Service applications and played a key catalyst role in the development of Service I&E geospatial data policy and repositories. In fact, this one initiative to share data had a much wider effect than anyone had anticipated. Because of these reasons, development of IVT data represents a useful case study and model for future I&E geospatial data-sharing initiatives within DoD, as will be discussed more in Chapter Eight.

In this chapter, we first discuss how the IVT data were developed, since that process is a useful model for future data development by the Services and the DISDI Office for DoD data sharing. Then we discuss how the data were used in the BRAC process. Last, we discuss some of the effects of the IVT data and process beyond BRAC.

Development of the IVT Data and Viewer Application

Since the IVT data had to meet BRAC auditing standards to ensure that quality data were being used, the development was a highly structured, highly controlled, and well-documented process. It is important to understand this process, because it provides a useful model for how OSD can acquire consistent and quality I&E geospatial data assets from all four Service installations for sharing. Here, we give a brief overview of this process by first describing what the IVT data consist of and how they were developed.

IVT Data and Viewer Application

The IVT process developed two main I&E geospatial data assets: IVT data and an IVT Viewer. In addition, the IVT also included extensive metadata to ensure that users understood the type and quality of the data. We describe each of these three components of the IVT.

IVT Data. IVT data consist of the GIS data layers and imagery for all 354 sites, meeting BRAC 2005 threshold criteria, including ranges. These data are summarized in Table 5.1. IVT data included installation overlays, which are basic detailed GIS datasets for each installation; one- and five-meter imagery for the installations; and additional data needed to examine the detailed installation data in regional and national contexts. The additional data layers consist of other key federal datasets to help with visualization and to provide additional information.

The installation overlays were acquired from each installation and the imagery data were acquired from a commercial firm. Most of the additional data layers were acquired from the USGS's National Atlas. However, the air-quality non-attainment areas were acquired from the EPA and the special-use airspace, military training routes, and air-refueling routes were acquired from NGA.

Figure 5.1 shows a sample of IVT imagery and installation boundary for NAS Whidbey Island, Washington.

The IVT Viewer. The OSD IVT Viewer is a customized version of a GIS software system that provided the BRAC 2005 process a way

Table 5.1
IVT Data

Category of Data	Dataset
IVT installation overlay layers	Installation boundaries Range complex boundaries Noise contours >65 decibels (> 60 decibels in California) Clear zones and accident potential zones Explosive safety quantity distance (ESQD) arcs 100-year floodplains Wetlands
IVT imagery	1-meter resolution imagery for installations or installation cantonment areas 5-meter resolution imagery for range complexes
Additional data layers	State and county boundaries City points and metropolitan statistical area (MSA) polygons Roads and railroads Hydrology Air-quality non-attainment areas Special-use airspace and military training routes Air-refueling routes

to view imagery and the GIS data layers in a consistent fashion for 354 sites. This tool was created for situational awareness only.

The system was designed to be user friendly and by people with limited or no GIS experience. The IVT Viewer was customized by shutting off some of the menu bars in the GIS software application that are used to perform more sophisticated GIS analyses. A specialized menu bar for IVT purposes was created. For example, it includes a special bar to see metadata. A user can click on the metadata button to acquire the detailed metadata, which includes the official signature pages. The system is organized around bases, so it is easy to view the different layers by choosing different bases.

The DISDI Office placed the IVT Viewer and underlying IVT data on seven laptops with 300 gigabyte external drives to hold the 233 gigabytes of data. Each IVT laptop system went to one of the seven BRAC JCSGs as a tool to help in their BRAC decisionmaking processes. Each group received some basic training in how to use the IVT Viewer.

IVT Metadata and Quality Assurance Plan (QAP). The IVT process included a well develop quality assurance (QA) and quality control (QC) process. Each IVT data layer had extensive metadata,

Figure 5.1
Sample IVT Map for NAS Whidbey Island, Washington

NAS Whidbey Island, WA

1 Miles   ☐ Installation Boundary

SOURCE: Map courtesy of the DISDI Office, 2006.
RAND *MG552-5.1*

which followed FGDC metadata standards and had been signed off
by official data stewards and installation commanders. These metadata, which included these official signature pages, were specified by an
extensive QAP developed by the IVT office in consultation with each
Service headquarter GIO to make sure that the process met both OSD
and Service needs. The QAP also required that the GIS vector datasets
meet the SDSFIE.

The IVT development process had an extensive QAP to ensure
that the highest quality data available were included and to make sure
that users understood the quality of the data they were using. All IVT
data had to meet BRAC auditing standards, which also contributed to
the extensive quality assurance process.

Each data layer had an official data steward who validated its quality with a signature. Installation commanders also had to validate their installation's data with their signatures. Each installation was required to explain if some data were not provided, such as if the data were not applicable for some installations. All this information, including the signatures, are part of the official IVT metadata. This extensive QAP and strict metadata development process enabled the IVT data to qualify as official OSD data that met strict and consistent quality standards across all installations. As will be discussed below, this QAP process became a useful model for the Services and also made the data more useful to users outside the BRAC process.

The IVT Development Process

Since this was the first systematic large-scale attempt to collect consistent I&E geospatial data across all four Services and classify the data as official OSD geospatial data, it is important to understand the data collection and development process. The IVT process began in August 2003 when the IVT program office (there was no official DISDI Office until the summer of 2004) started writing the quality assurance plan. In September/October 2003, the Services started delivering their data.

The basic installation data for IVT were provided by each installation to the IVT program office and conformed to the IVT QAP and extensive standards guidance. Each installation submitted a separate CD of data. The Services headquarters geospatial information offices acted as facilitators in working with the regional offices or individual installations to supply the data and perform QA/QC in the process. The IVT program office worked with the installations and the headquarters facilitators as problems arose, such as cleaning up metadata. Service IVT coordinators and mission knowledge experts helped in this process. USAF auditors also checked metadata during the development process.

In June 2004, the first version of IVT was delivered to OSD BRAC decisionmakers. In August 2004, the IVT program office delivered an updated version of IVT with complete metadata. The Navy BRAC teams requested sea training ranges, which had not been in the original system, so this update was made and the final IVT system was delivered in October 2004.

The official IVT data and Viewer from this process include data that have not been updated and are from fall 2003 if not earlier.

IVT Data and Viewer Application Use in the BRAC Process

The IVT data and Viewer were designed to provide situational awareness for decisionmakers in the BRAC process. It is important to note that no analysis was done with the IVT Viewer itself. However, IVT data were used to support other analyses. Four key decisionmaking groups in the BRAC process received and used some form of the IVT data or the Viewer in their BRAC decisionmaking process:

- the JCSGs
- the Service BRAC offices
- OSD leaders
- Congress.

We investigated how decisionmakers in each group used the IVT and discuss these uses here.

JCSG Use of IVT

The IVT Viewer Tool on an independent laptop computer with an external hard drive went to each of the seven JCSGs. The groups were

- Medical
- Intelligence
- Education and Training
- Headquarters and Support Activities
- Industrial
- Supply and Storage
- Technical.

We interviewed members of the Education and Training, Medical, and Industrial JCSGs. Those in the Industrial JCSG did not use the IVT Tool very much because they had few installations to examine and had extensive knowledge of the geography of their bases. Also, their installations tended to be smaller and more urbanized and, thereby, did not need to use geospatial information. For example, geospatial issues such as accident potential zones and training range locations are not relevant at such installations. However, the Medical and Education and Training JCSGs both used IVT data to support their decision-making processes but in very different ways. We discuss some of their sample uses here.

Education and Training JCSG IVT Use. The Education and Training JCSG consisted of four subgroups: the ranges, flight training, specialize skill training, and professional development education and training subgroup.

The flight training subgroup used the IVT as it was designed to help provide situational awareness. When examining an installation, group members would bring up the installation data to look at various issues, such as any constraints to or encroachment on flight clear zones or approach zones. The IVT Viewer was especially useful for looking at accident potential zones (APZs) and clear zones around airfields, which was a main concern for this subgroup. Also examined was the amount of built-up area around a base and the uninhabited part to learn whether the areas were not built up because of constraints on the acreage. The IVT Viewer helped this subgroup understand the context and provide an overview of key geographic issues at a base, such as encroachment factors. As one JCSG member stated, we used it to "give us a warm fuzzy on what we were dealing with."

The ranges subgroup combined the IVT data with other geospatial data to support other analyses and to create a more powerful analytical tool. This subgroup included GIS analysts who combined the IVT data with additional GIS range data from the Sustainable Range Program and with other federal agency data, such as other federal land locations, for the Bureau of Indian Affairs, the Bureau of Land Management, the National Park Service, and the Forest Service. The group added other analytical features, such as showing 50-, 100-, 150-, and 200-mile buffers around key locations. These data were used to support different analyses including assessing the feasible locations for urban operations training, analyzing range relationships and restrictions, and helping perform the BRAC military value calculation.

This subgroup was tasked with identifying and making recommendations about where to place some additional urban operations training. By combining the IVT and other GIS data to show key characteristics, such as size, relationships to special-use airspace, other federal lands, etc., group members were able to develop maps to show where and how it was feasible to conduct urban operations training. PowerPoint slides of the maps were created and used to brief flag officers so that they could see how urban operations could be conducted. It helped senior leaders more quickly and better understand what the JCSG was proposing and what the implications were. Even after the official BRAC decisions had been made, these slides were still being used to show senior managers about the new locations for urban operations training.

The IVT and other GIS data were combined with BRAC data call[1] data and used to support the analysis of installation range relationships and general restrictions. The BRAC range data call included detailed training range information, such as restrictions on individual ranges (i.e., threatened and endangered species, cultural resources, and wetlands). The geospatial information helped managers examine key geospatial relationships, such as which ranges were contiguous, how big they were, and where the different general restrictions were and how much of the range they affected. GIS data were also useful in helping develop key questions about relationships and in requesting additional data from the installations in the official BRAC data calls.

The ranges subgroup geospatial analysts used this geospatial information to help perform the BRAC military value calculation. Part of the range military value included looking at the proximity of air and ground ranges. GIS data were used to analyze which bases were within a fixed distance of which ranges. Using IVT and other geospatial data, the mileage from bases to the ranges was calculated. The results were linked with information from the BRAC data call, i.e., range charac-

---

[1] The 2005 BRAC process involved official data calls to acquire consistent, reliable data about installations.

teristics, such as whether laser operations were possible. These GIS-created data were used in the distance equation for determining military value.

The IVT data became part of a more powerful geospatial analysis and display tool, because this subgroup had additional geospatial data and analytical support. In summary, having IVT combined with other geospatial data and technical assistance helped the ranges subgroup perform its BRAC analysis, helped it develop key questions, and helped in briefing high-level decisionmakers who understood things better because they could see what was going on. It helped them improve their decisionmaking process and saved them time. In fact, one JCSG subgroup member stated that it would have been "more difficult to do the mission" without the geospatial data. He also stated that every time they briefed someone, and there were many briefings, they had 20–40 PowerPoint charts and about 40 percent of them were maps that had used the IVT and other GIS data.

Medical JCSG IVT Use. The Medical JCSG used IVT data to create large maps for examining issues of concern. It used these maps to help support other analyses, including siting medical facilities at installations, checking to see if there was space available at installations, and helping to convince the Service BRAC offices of the best location for medical facilities.

Once the Medical JCSG determined that it needed new medical facilities at a specific installation, siting them on the installation became an important analysis, coordination, and communication issue with the relevant Service. This is because all the JCSGs had to coordinate with the Service BRAC offices in assessing where to place new facilities on an installation to ensure that there was space available and that the space was not already designated for some other purpose. Obviously, not everyone could build something on the same piece of land. In addition, the cost and feasibility of a given scenario was affected by whether the new facility was located in existing or new buildings.

In one case, the Medical JCSG used the IVT maps to help support an analysis of where to site a new large medical facility. The group printed out a large map of the IVT imagery data, boundaries, and roads for the installation. Then, group members looked at the open space and location of key buildings on base. Using this map and other information about the base, such as which functions were in which buildings, the best place to locate the medical facility seemed to be on some open space near the installation commissary. This IVT map was used to work with the Service BRAC office to communicate and justify the need for this piece of land. The IVT data helped the group more effectively negotiate, work, and communicate with the Service office to be able to use the desired piece of land.

In another example at a different installation, the Medical JCSG proposed building a new clinic, which required 40 acres. The Service BRAC office staff members said that they had only one-half acre for such a facility, judging by their review of the base facility plan. The Medical JCSG thought that there was quite a bit more available acreage at the base. Using IVT data to support an analysis process and to show many feasible locations for such a clinic, group members communicated with the Service office, which did a more in-depth analysis of what was planned at the base and what the true available space was. The Service office realized that 300 acres were available. The maps created with IVT data had helped serve as a useful check on available acreage at the base and also helped improve the Medical JCSG communication process with the Service office.

There were other IVT application examples, with similar benefits. To summarize, the Medical JCSG found that by using IVT data, decisionmaking processes were improved, such as helping with the siting analysis and decisions about locating new medical facilities at bases and checking available acreage. They also helped give the military medical leaders the confidence that if they changed a siting decision, there would still be enough acreage available elsewhere at the base. The data made the Medical JCSG more comfortable with the group's recommendations and improved communication and coordination with the Service BRAC offices. In addition, the Medical JCSG was able to make siting recommendations for the bases in about half the time because of the IVT data.

Even though the Medical JCSG found IVT quite useful, one member commented how it could have been a much more useful tool with some additional data and analytical functionality built into the system. Specifically, the system could have been used more effectively if it had been designed more to look at medical concerns, such as including population densities and military treatment facility (MTF) capacities in the system. Medical decisions involve knowing where the populations are and how far people have to drive to reach medical care. One JCSG member wanted this expanded GIS capability on his desktop, but it was too time-consuming and expensive for the Medical JCSG staff to add in the data and analytical capabilities that he wanted. It is important to note that the IVT was not designed as such an analysis tool, but this is useful feedback for any future efforts to develop such capabilities for supporting an OSD process such as another BRAC round. It also illustrates something that we learned in talking with other decisionmakers when they are first exposed to I&E geospatial data assets. They often want more than just situational awareness capabilities; they want more analytical capability.

Service BRAC Office Use of IVT

IVT data, but not the IVT tool, were delivered to each Service's BRAC office. The Services found these IVT data to be useful in their BRAC processes and here we discuss how the USAF and U.S. Army BRAC offices used the data.

IVT Uses by the USAF BRAC Office. The USAF BRAC office combined the IVT data with other geospatial data and used them in their BRAC analysis, communication, and outreach processes. This office had GIS staff to help use the IVT data.

The Secretary of the Air Force appointed a Base Closure Executive Group (BCEG) of six general officers and seven comparable (Senior Executive Service) civilians. Additionally, an Air Staff-level Base Closure Working Group was formed to provide staff support and additional detailed expertise for the Executive Group.

The USAF BRAC office GIS staff produced the "Base Closure Executive Group (BCEG)–BRAC Reference Book," which is an 11-1/2-inch by 20-inch map book. It contains 14 maps, which used IVT data combined with some additional geospatial data. The maps cover the entire United States, unless otherwise noted, and include

- installations, ranges, special-use airspace, and military training routes (MTRs) for the entire country and for the northwest, northeast, southwest, and southeast CONUS and for Alaska, Hawaii, Puerto Rico, and Guam
- population density and military installation locations
- federal lands, installation and range locations
- commercial air traffic air tracks on Thanksgiving day, 2003, for 8:00–17:00 Eastern Standard Time (EST) shown with special-use airspace
- commercial air traffic air tracks on October 16, 2003, from 8:00–17:00 EST shown with special-use airspace
- military air tracks on October 16, 2003, from 8:00–17:00 EST shown with special-use airspace
- general aviation tracks on October 16, 2003, from 8:00–17:00 EST shown with special-use airspace
- areas in air-quality non-attainment for 2003, installation and range locations
- areas in air-quality non-attainment any time between 1992 and 2003, installation and range locations.

This book was given to the 12 members of the USAF BCEG and the USAF BRAC Commission. The latter group also shared it with members of Congress and other Service BRAC offices as part of its BRAC outreach and communication process.

The BCEG maps and IVT data combined with other geospatial information were consulted during the USAF BRAC commission decisionmaking. The data were useful for supporting the analysis and examination of aircraft beddown options and in helping examine and explain operational encroachment. For example, for one base near a small western town, the IVT data were used to show that this installation was "operationally encroached," because the town is on a main commercial air flight route. This is obvious from the map, which shows Federal Aviation Administration (FAA) commercial air traffic air tracks overlaid with the IVT special-use airspaces near the installation. See Figure 5.2 for a sample of such commercial air traffic air tracks from October 16, 2003, from 8:00–18:00 EST.

Figure 5.2
Commercial Air Traffic Air Tracks on October 16, 2003

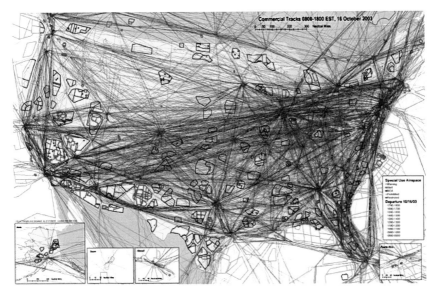

SOURCE: Map courtesy of HQ AF Ranges and Airspace, July 28, 2005.
RAND *MG552-5.2*

Senior USAF BRAC staff used IVT combined with other geospatial information "to make a point" to JCSGs, senior military leaders, Congress, and others, for example, to show why they proposed removing aircraft that were stationed at a base near a central midwestern city. The map that includes FAA commercial air traffic air tracks and IVT special-use airspaces shows immediately all the commercial air traffic lines over that city and the resultant congestion. With the map, the senior decisionmakers did not need to see the USAF's analysis process for this decision. As one USAF staff member stated, they also did not need to ask questions.

IVT Uses by the U.S. Army BRAC Office. The U.S. Army BRAC office, called The Army Basing Study (TABS) office, combined IVT data with other geospatial data to help in their analysis process and to help provide a visual installation orientation at TABS installation orientation sessions.

For each of about 100 Army installations (all the IVT bases), the TABS office received a briefing by the base commander and base staff, which was a complete overview of the installation, such as the mission and interaction with the community. ACSIM geospatial staff set up a laptop-based interactive GIS display of each installation using the IVT data that was projected on the wall during each presentation so that TABS staff could visualize the installation and see the specific locations that were being talked about. The display had been coordinated ahead of time so that the staff member could zoom into the image of whatever area on base was being talked about. These displays helped the analysts who had not been to the base better understand and visualize the base. The installations also brought hard copy maps of the master plan for the installation.

ACSIM geospatial staff provided a number of other analytical products or analyses to the TABS office using IVT data to support these other analyses. Here is a sample of such items:

- a special ARCReader product of IVT data for TABS analysts to use at their desktops so that they could easily bring the data up at any time
- for selected installations, IVT imagery and noise zones combined with census population data to help anticipate noise issues that might arise at those installations
- for a limited set of installations that TABS identified as potential gaining installations, installation maps with "land-use control" information and with one- and five-meter grids to help estimate the number of "buildable acres"
- a hard copy portable map book with all the IVT data layers for each installation, so that TABS staff could use it easily at different meetings
- several large wall maps of the IVT imagery and installation boundaries combined with other geospatial data, such as place names, to help TABS staff familiarize themselves with these bases.

The USACE CERL also provided geospatial analysis to TABS in which they used IVT data. IVT imagery and base boundaries were combined with other geospatial information to analyze encroachment from urban sprawl around 100 Army bases. This analytical process is called the land-use change analysis (LUCA).[2] The information from this analysis was also used in the "future options" of the BRAC military value calculation. Having certified and consistent data for 100 installations was critical to the USACE analysis, which could not have been performed without the IVT data.

OSD Leadership IVT Uses

The OSD BRAC office requested a series of special map products from the DISDI Office for the Secretary of Defense (SECDEF) and other senior OSD staff between February 18, 2005, and May 20, 2005. IVT staff produced and delivered about 350 maps during this time period. The DISDI staff received about two to three requests a week. A staff member from the OSD BRAC office would call the DISDI staff and ask for specific types of maps for an upcoming briefing. Maps were delivered as JPEG files. The expected turnaround time to produce and deliver such maps was around an hour. After the first few requests, DISDI staff began to anticipate calls and prepared some of the maps ahead of time. Given the tight time constraints, the DISDI staff could not have responded to such requests without the IVT data.

These maps showed "installation dots" by state, region, or the entire United States that represented "closures," "realignments," and "gainers." For example, one map showed major closures and another major realignments for the entire country, whereas others showed the dots of closures, realignments, and gainers by state. See Figure 5.3 for a sample of this type of map for Missouri.

Senior OSD leaders found these maps useful as they reviewed the BRAC recommendations. The maps helped illuminate the discussion about different areas of concern and conditions that might not otherwise have been apparent. In one case, one map helped leaders change a decision after they realized that a significant number of people within the region would be affected by a potential closure. Insights from reviewing

---

[2] For more information on this process, see Lozar et al. (2005).

Figure 5.3
State Installation Dot Map for Missouri

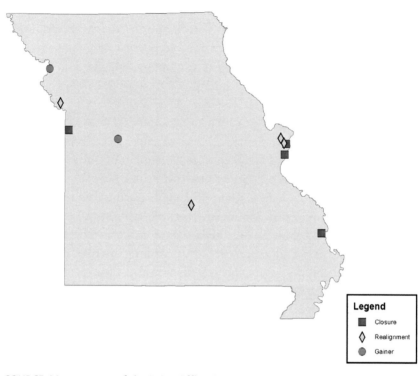

SOURCE: Map courtesy of the DISDI Office, 2006.
RAND MG552-5.3

the map resulted in the proposal being reevaluated and readjusted to lessen the effect. These maps and the OSD BRAC staff's additional use of the IVT tool helped to save senior staff the time and cost of visiting the bases.

The Presidential BRAC Commission and Congressional IVT Uses

The Presidential BRAC Commission and Congress also used the IVT data to help review BRAC decisions. In June 2005, special PDF map products were delivered by the DISDI Office to the OSD BRAC office for Congress and the Presidential BRAC Commission. Because information going to Congress could go to the public, OSD IVT data were modified because of sensitivities within the data that made some information unsuitable to be available in the public domain. There were four modifications:

- ESQD arcs were removed because of security concerns.
- Government installation points of contact in the metadata were removed.
- The disclaimers were removed from the metadata.
- IKONOS imagery data files were removed because of commercial licensing restrictions.

The PDF map products consisted of one map of the imagery for the base, one map with the installation overlays on it, and the metadata data files for all 354 installations/ranges. Figure 5.1 shows one of these products.

Another product using IVT data also was used by Congress. As mentioned above, USAF BCEG maps helped members of Congress understand flight "operational encroachment" and other issues that affected USAF BRAC decisions.

Diverse Value from IVT Use in BRAC

The IVT data, when combined with other geospatial information and in supporting other analyses, provided more than just "situational awareness" in the BRAC process. There were diverse other uses: the Medical JCSG used the data to help support analyses of where to site medical facilities at installations. The Education and Training JCSG used them to help support the military value calculation by the ranges subgroup and provide situational awareness for the flight training subgroup. The USAF BRAC office used them to examine aircraft bed-down decisions and operational encroachment from commercial air traffic. IVT data enabled the Army BRAC office to visualize bases without having to visit them and to help assess urban sprawl around bases. IVT data helped explain BRAC decisions to senior leaders and Congress and also helped in BRAC implementation, as discussed in examples in the appendix.

To understand the value of having, using, and sharing geospatial information in the BRAC process, we summarize some decisionmaker interviewees' comments about how IVT improved the decisionmaking process:

- Points out things we had not noticed before
- Causes you to ask questions
- Could see what is happening better
- Used to help make a point
- Clarifies things
- Helps visualize gut feelings
- Provides quick information in an intuitive fashion
- Provides more confidence in the decisions
- Eases communications
- Enables more complete solutions
- Is useful for common situational awareness.

Key Value Added Benefits of IVT in BRAC

In assessing the use of IVT in the BRAC process, some common value-added benefits were seen across the different uses. First, IVT data enabled different BRAC staff members to quickly produce and generate numerous maps and PowerPoint charts. Given the tight time deadlines of the BRAC process, this was an important contribution; much of this work could not have been done in other ways because of time constraints. Second, IVT data helped provide the ability to integrate installation data with information outside the installation and assess relationships, such as ground and air space encroachment analysis. The data also helped to show relationships with local community, public, and congressional concerns. Third, IVT data provided a key communication tool and supported other analyses. The data had an important role in helping explain the BRAC decisions to senior decisionmakers and others, as well as in implementing BRAC recommendations. In addition, the official common OSD data source made the data more valuable, especially for joint considerations across multiple Services.

Other Effects and Uses of the IVT Data and Process

In our assessment of the BRAC use of IVT and during interviews within other I&E geospatial communities, we discovered many additional uses of IVT data and effects from the IVT process. IVT helped generate interest in other DoD and non-DoD organizations that wanted to use such information, increasing personnel's knowledge about the data and their potential future uses. The IVT process itself, especially the QAP process, is a model that has been used by the Services to help develop their I&E geospatial programs. We briefly discuss some of these effects here. For discussion purposes, they are grouped into five areas:

- Service headquarters uses of IVT data
- other Service use of IVT data
- other DoD and non-DoD uses of IVT data
- IVT data as a foundation for DoD geospatial data portals/ repositories
- IVT process as a useful model for the Services.

Service Headquarters Uses of IVT Data

The IVT data provided a consistent set of installation geospatial data layers across multiple installations. Because the data met strict OSD quality standards, they provided validated data at the Service head-quarters level that had not existed before. The Service headquarters offices, especially the Service GIOs, have used these data in various ways. We illustrate this point with some uses by the U.S. Air Force and Army GIOs.

Senior USAF leaders need authoritative data on demand. Having the IVT data has made that type of authoritative data available to them. The HAF GIO now uses these data to support senior leaders' needs in a timely fashion. For example, it used IVT data to

- provide data to the USAF Installations and Logistics Crisis Action Team for Hurricanes Katrina and Rita

- assess whether the Department of Energy's (DOE's) placement of energy corridors in the western United States would touch any USAF installation properties
- provide maps in the "trip books" (information about the base) for the USAF Chief of Staff and other senior staff making trips to installations.

The Army GIO office staff at ACSIM has also used IVT data to analyze and produce maps for senior leaders. For example, staff members have supported Army transformation master planning by supplying detailed maps. For this task, they created a series of maps, such as one of the "Fort Carson NW, Colorado Transformation Master Plan," to show where construction would take place. This map uses IVT imagery and GISR roads data. Map details include information about new construction, existing facilities, relocatable facilities, and renovation/conversion. The maps are supplied in JPEG format for PowerPoint briefings and in PDF files for printing (gave E-Size plots for large printing).

Army GIO staff members also used IVT data to help answer requests they receive from other DoD organizations for geospatial information. For example, the Pentagon Legislative Affairs, Office of Chief Legislative, wanted 50 state maps showing the overlap between installation boundaries and legislators' jurisdictions. Staff members used Army IVT data to help produce these maps.

As will be discussed more below, the use of IVT data has helped give senior leaders a taste of the usefulness of I&E geospatial data. For example, the Chief of Staff of the Air Force carries the BCEG map that shows FAA commercial air traffic air tracks and IVT special-use airspaces with his set of favorite briefing slides. He finds this map useful, especially in helping to explain operational encroachment.

Other Service Use of IVT Data

Since IVT data originated in 2003 at the installation level and installations have the most recent version of such data, most of the installations did not use the older IVT data. However, the exception is imagery data, which many installations themselves do not have. IVT data were most useful to Service organizations examining multiple sites, such as Major Commands, functional areas, and regions. Here, the IVT data were used by other parts of the Services to help in various analyses that cut across installations. For example, as discussed above, the USACE CERL used IVT data, especially imagery, in its encroachment analysis during the BRAC process. The data have also been used to help the USACE help Army installations examine conservation buffers and sustainability issues.

USAF planners have also found IVT useful when making non-BRAC aircraft beddown decisions. For example, at a northern midwest base, they used a map that shows FAA commercial air traffic air tracks overlaid with the IVT special-use airspaces (see Figure 5.2) to show that because of commercial air traffic congestion, it was not a good decision to bed down a new U.S. aircraft at that base.

The Air National Guard (ANG) has used IVT data to help with parking plan analysis. ANG civil engineers used ANG IVT data to examine ANG bases that might be candidates for change and to determine the effect on the bases. ANG engineers mainly used one-meter imagery data from IVT data to examine aircraft parking issues. They used IVT data for 20 to 50 bases. For some smaller bases not in IVT, they used USGS one-meter data (from seamless.usgs.gov).

Other DoD and Non-DoD Uses of IVT Data

Because IVT data were used in the BRAC process and the DISDI and Service portals and were being discussed by DISDI staff at conferences, much interest in and requests for IVT data were generated. Other parts of DoD, such as OSD offices and NGA, and non-DoD organizations, such as the USGS and U.S. Coast Guard, have been interested in acquiring and using IVT data. Such organizations value IVT data because they meet strict QA/QC criteria set by OSD. For example, the fact that installation commanders signed off on the data made the data more valuable to users, since they had been validated by this authoritative source.

IVT data have helped to create awareness and to generate interest in I&E geospatial data assets across many diverse DoD organizations. The DISDI Office has received numerous inquiries by other

OSD offices that want to use IVT data to support their activities. Such organizations have included the DoD Explosives Safety Board, the Office of the Assistant Secretary of Defense for Homeland Defense (OASD(HD)) Program Office for Mission Assurance, the Defense Critical Infrastructure Program (DCIP), the OSD Health Affairs TRI-CARE[3] Management Activity (TMA)/Health Programs Analysis and Evaluation Directorate, the OSD ESOH staff, and the Defense Threat Reduction Agency. We have already discussed several examples of how these organizations want to or have used IVT data, such as the OSD ESOH staff using DISDI's analysis and maps with IVT and other geospatial data to identify focus areas for SERPPAS. Another example is that the OSD Health Affairs TMA/Health Programs Analysis and Evaluation Directorate has been developing a "Military Health System Atlas" to help examine and assess military medical capabilities and their populations; they requested and received IVT data for this task.

NGA has also requested and received IVT data for use in Palanterra, its web-viewing system for U.S. geospatial information to support homeland defense and security.

Non-DoD organizations, including federal agencies, states, and NGOs, have also requested and received IVT data from DISDI. For example, conservation NGOs, such as the Nature Conservancy and the Conservation Fund, have requested IVT data to help in the development of conservation easements as buffers around military installations.

IVT Data as a Foundation for DoD Geospatial Data Portals/ Repositories

The IVT data provided a consistent basic set of installation geospatial data layers that met strict QA/QC standards across multiple installations, providing a fundamental data capability at the OSD and Service headquarters level that had not existed before. IVT data have therefore become a starting foundation for OSD and Service-wide I&E geospatial data repositories and portals. The DISDI Portal uses IVT imag-

---

[3] TRICARE is the health insurance plan for U.S. military members, their families, and military retirees.

ery and GIS vector data layers as its basic set of I&E geospatial data for the United States. The Navy also uses the basic IVT imagery and GIS vector data layers for Navy installations as the initial foundation of the GeoReadiness repository. Similarly, the Army uses the IVT datasets for Army installations as the standardized official foundation data for the Army's GISR. However, unlike DISDI, the Services have been updating these data. For example, the Army has incorporated in GISR updates from installations on some of the vector layers, such as installation boundaries. The Services and DISDI have both added other data to their portals, for example, both the DISDI Portal and GISR include some additional transportation data, such as some data from the NAVTEQ Company.

IVT Process as a Useful Model for the Services

The IVT process was used by the Services in two key ways: It made the process more visible to senior leaders and led to more high-level support. Furthermore, the QA/QC process has been used as a basis for the Service QA/QC processes. We discuss each below. In addition, the Army used the organizational structure from the IVT process as a basis for its I&E geospatial organizational structure in a new Army regulation about I&E geospatial data.

The IVT process gave all the Services more visibility for their I&E geospatial data programs. It was a key catalyst in furthering the development of these programs. It gave senior headquarters and installation leaders a view of what geospatial data could do and it increased awareness about data availability and usefulness. In some cases, this led to more support and development of the Service geospatial programs. For example, the IVT process enabled the Army to establish regional coordinators for geospatial information within IMA, called the "IMA regional GIS managers." In addition, the Army was able to leverage funding from the IVT data to help develop the GISR. The USMC also leveraged off the IVT data-collection process to help develop and collect additional I&E geospatial data assets. When traveling to installations to gather IVT data, USMC staff also collected other geospatial data to include in the GEOFidelis Portal. Similarly, the IVT process helped the USAF HAF GIO receive more headquarters support and management approval to develop the entire GeoBase program.

Since installation commanders had to review and sign off on their installation geospatial data as part of the official OSD process, some commanders, who had not been aware of their geospatial program's capabilities, began to appreciate the usefulness and significance of such data.

The IVT QAP and entire QA/QC process was a useful model for all four Services. The Navy has used the QA/QC process from the IVT to help develop its GeoReadiness repository. The USAF used the IVT QA/QC process as a template for developing its Common Installation Picture QA/QC process. In fact, the Navy finds that the process gives a useful template when developing other geospatial data as well, i.e., both CIP and mission datasets. The U.S. Army is in the process of institutionalizing the IVT data-development process, including the QAP, for Army I&E geospatial datasets.

Summary of the Effect of the IVT Data and Process

IVT data helped improve the BRAC decisionmaking process and saved time and money. It helped point out things that would not have been noticed otherwise. IVT data were used to support other analyses, such as military value calculations. They helped explain BRAC choices to senior leaders, such as generals and Congress members.

The use of IVT data and the IVT Viewer provided information that could not have been acquired in time in other ways. As a stand-alone tool, it provided situational awareness. When the IVT data were combined with other geospatial information and support, the information became part of a powerful analytical tool that was used to support other analyses in the BRAC process.

Many unanticipated spin-off benefits resulted from both the IVT process and the data itelf. IVT was used for many purposes outside BRAC. The strong quality assurance process was an important part of acceptance and use of the data, both within and outside the BRAC process. First, the Services themselves found the IVT data useful outside the BRAC process. The data provide the Service headquarters office with an authoritative multi-installation dataset that is used to support senior Service decisionmakers. Functional and regional organizations also took advantage of this capability in analyses looking across multiple installations. The IVT data also were a catalyst and foundation for Services developing Service-wide data portals and repositories. Second, the DISDI Office used IVT data as a foundation for its DISDI Portal. The office also used the data for outreach and to educate others about the availability and usefulness of I&E geospatial data assets. Third, other parts of DoD, such as other OSD offices and NGA, have found the data to be a useful resource in their mission activities. Fourth, organizations outside DoD, such as other federal agencies, state agencies, and NGOs, are using IVT data in their activities in such areas as homeland defense and environmental management. Fifth, the QA/QC process has been used as a basis for the Services QA/QC processes in developing their I&E geospatial data and ensuring consistent high-quality data across the Services. Last, the IVT process has made Service I&E geospatial programs more visible to senior leaders and the ensuing support has helped in the continuation of these programs.

CHAPTER SIX

Future Use and Sharing of I&E Geospatial Data Assets

In Chapters Three and Four and the appendix, over 150 examples are presented from diverse mission areas showing the widespread use of I&E geospatial data assets. Our analysis shows that such trends are likely to continue into the future and that the use of I&E geospatial data assets to support mission applications will continue and also expand to other mission areas. In this chapter, we discuss the likely implications for future I&E geospatial data asset sharing.

Three key areas are addressed. First, we discuss how the trends we see suggest that demand and use of I&E geospatial data assets by different organizations both within DoD and outside DoD are likely to increase. Second, we discuss the current barriers to I&E geospatial data asset sharing. These barriers are important to understand because if they are not addressed, they could limit potential future sharing. In addition, the current evolution of technology makes it even more important to address these barriers. Thus, third, we discuss some changes that might occur in how the assets are used as suggested by evolving technology and market applications.

Increasing Demand and Use of I&E Geospatial Data Assets

In previous chapters, the discussion focused mostly on current mission application examples, since there are so many of them. In the future, the trend appears to be that more applications will be found within and across all levels—base, major and functional commands, headquarters, and OSD. This continued spreading of the use and sharing of I&E geospatial data assets results for a number of reasons: First the data and technology are now easier to use in more user friendly ways, such as in web-based systems; second, standards and interoperability conditions are being implemented that help facilitate multiple use and sharing; third, efficiency and effectiveness benefits are being realized, which helps facilitate investment in these resources; fourth, sharing is mandated by OMB Circular A-16; and fifth, centralized military organizations, such as the Service headquarters offices and DISDI, are helping to facilitate the use and sharing of such assets.

Given these factors and the continued evolution of the geospatial technologies themselves, some growth trends are likely for the future mission applications, which were illustrated by some of the examples discussed in other parts of this monograph.

More Use by the Warfighter and the Intelligence Communities

Because of the synergies in sharing geospatial data, applications, techniques, knowledge, and skills, I&E geospatial data assets are likely to become more useful to the warfighting and intelligence communities. In Chapter Four, we discussed how these communities have already begun using I&E geospatial data assets to help in C4 systems; in combat and post-conflict operations; in mobilizations and deployments; in supporting base camps and other forward operating sites; in antiterrorism and emergency response activities; in logistic operations; and for warfighting planning, wargaming, and assessments. Such uses should continue to grow.

More important, there is a changing relationship between geospatial operations at the installation level and the warfighting and intelligence communities. Historically, these two communities have not interacted much. However, because of increasing synergies between their missions, as discussed in Chapter Four, this has begun to change. Key barriers to such collaboration still exist, including the different training, orientation, culture, and stovepipes of these communities and security issues. Another barrier is that many in the warfighting and intelligence communities do not know about the existence of I&E geospatial data assets. However, this new relationship will evolve more in the future because of the benefits in collaborating to improve the speed and effectiveness of the U.S. military's ability to rapidly deploy and respond where needed around the world, whether to fight the Global War on Terrorism or to provide humanitarian assistance.

A good example of how such an evolution—although a difficult one—relates to the MIM production collaboration process. As discussed in the appendix, a MIM is an installation map that warfighters use for training. NGA used to develop, certify, and produce these maps but does not want to produce them any longer.

DISDI is working to have the installations produce these maps with NGA still having responsibility to officially certify them and mass-produce the hard copies. In 2005, in a pilot experiment, NGA worked with Camp LeJeune to develop the map locally by installation GIS staff. This experiment is helping to work out the logistical and process issues involved in having installation staff collaborate with NGA. If this experiment works, DISDI and NGA will try to change the way these maps are produced.

To illustrate the complexities and challenges of such collaborations, we briefly discuss some advantages and disadvantages to this new collaborative process:

- increases timeliness of the data in the product
- has a quality effect that will most likely vary by installation; could be better or worse depending on the individual installation GIS skills and data quality (in most cases, most of the MIMs are expected to be of better quality, because the most up-to-date data are being used from the source at the installation; quality would likely improve over time at the installations that lag behind)
- saves NGA manpower, time, and some funding
- brings more business to installation GIS staff, causing extra workload and, thereby, is an additional cost; but staff may receive more money or technical assistance to perform this task

- could help some installations meet some minimal GIS skill level needed to perform this function, especially if assistance was received.

Over the long run, collaboration would improve the efficiency and effectiveness of producing MIMs, most importantly by providing more accurate and more timely maps for warfighting training. It would also help address some problems that have occurred because the Army's "Fort X Special" maps[1] and MIMs are both being used in the field. For example, in a training exercise at Fort Hood, some soldiers were using a more up-to-date "Fort X Special" map than others. Unfortunately, the two maps had different training area numbering schemes. Luckily, the difference in maps was identified before the live fire training began, which could have caused many communication problems and even some injuries. This example also illustrates the importance of having more communication and coordination between the installation and intelligence and warfighting geospatial communities.

Many people whom we surveyed in warfighting and intelligence operations thought that a common operating environment, with the same standards and formats, is needed for geospatial information for the I&E, warfighting, and intelligence communities. This would help these communities increase future use and sharing. There are benefits to more sharing of I&E geospatial data assets between the warfighting and intelligence communities. However, more interoperability between these communities is needed to make the sharing of data and applications efficient. To illustrate why this is so important, consider that currently with military operational deployments, different military organizations use different geospatial information management systems, often with different standards, software, data types, and formats. For example, there are I&E geospatial data systems at the installation, other systems used by the military transporters, and different

---

[1] A "Fort X Special" map product is an installation training map produced by the U.S. Army ITAM for installation training purposes. It was developed as substitute for a MIM. It is similar to an NGA MIM with the same scale and symbols. For more information, see the appendix.

geospatial systems used on the ground in the warfighting operation, such as those developed by NGA. If the U.S. military needs to rapidly deploy within 72 hours, having such different information systems, which often require data format and translation processes, slows down the process. Improving the interoperability of the geospatial management, analysis, and information systems will help improve military readiness, response times, and operations.

In fact, the USMC is exploring the concept of having an enterprise approach to geospatial information across the entire USMC because the warfighters, intelligence community, and installation geospatial staff and missions all benefit from sharing data assets and skills.

More Demand and Use by Other Parts of OSD and DoD

In the previous chapters, several examples were presented of other OSD offices that are using I&E geospatial data assets, such as environmental, health affairs, and explosive safety planning. In interviewing staff from such organizations, in examining other OSD activities, and in analyzing the application examples and potential future examples, it is clear that OSD organizations' interest and desire to use I&E geospatial data assets is likely to increase. The same is true in other parts of DoD management. OSD and other DoD management organizations have good business reasons for using more I&E geospatial data assets, which should help them being used more. For example, OSD and Service headquarters organizations and functional commands all can improve asset management and upward reporting by using shared I&E geospatial data assets. Integrating, aggregating, and sharing geospatial information from installations to higher management, in areas such as real property, environmental issues, military health capabilities, and safety, can help provide better quality and more consistent reports to improve decisionmaking and management processes. More of the reporting processes could be automated and save costs as well. Evidence of increasing interest and demand in I&E geospatial data by other parts of OSD is also seen with the Real Property Inventory process. DISDI is currently working with other parts of OSD to improve their RPI processes. RPI is a formal registry of accurate real property data, including site locations, giving DoD an accurate account of its properties. Mapping legal boundaries is a key part of this process. Using I&E geospatial data assets to develop a more accurate and geospatially enabled RPI will help DoD organizations better manage such properties.

Even some DoD organizations and mission areas, such as manpower and acquisition, that do not seem to have any use for I&E geospatial data assets could benefit from their use and in some cases already have. Any application that involves geospatial information, such as an address, often can benefit from geospatial analysis to better assess or manage resources. An interesting example to illustrate this point with respect to manpower concerns occurred with the Council for the Commandant of the Marine Corps (CMC). The CMC wanted to better understand who was doing what and where, by examining where USMC commands were, i.e., at which Navy and USMC installations. USMC GEOFidelis staff produced a map showing the locations of different commands and units throughout the world so that the CMC could better understand the functional manpower situation.

An acquisition example occurred with the Army acquisition community in its design and analysis of future combat systems. As part of the design process, staff members built a subset of vehicles and simulated the rest of them. In this simulation, they used I&E geospatial data, such as terrain and slope information, to help test the future combat system. Such a simulation is being run at Fort Bliss, Texas.

More I&E Geospatial Data Asset Use by Nonmilitary Communities and Increased Demand for Acquiring Nonmilitary Community Geospatial Data

Many DoD mission areas that use I&E geospatial data assets are ones that collaborate, partner, and interface with nonmilitary organizations. There is a large amount of current sharing outside DoD with other federal agencies and state and local governments and such sharing will likely increase given current trends. As discussed above, other U.S. government agencies need geospatial information to help with key functions, such as homeland security, environmental management, emergency response, and land-use planning. Often, the military works with other U.S. government agencies in such functions and collaboration is increasing. Numerous examples of such sharing are discussed in Chapter Three and the appendix.

In fact, many of the military geospatial data developers and users whom we interviewed were also very interested in sharing data outside DoD and felt that the issue of data sharing outside DoD should be a focus for our study. Many interviewees needed other organizations' data and wanted reciprocity agreements for data sharing. Military installations want and need access to local, state, and federal data to help perform their missions. For example, the USAF Academy in Colorado Springs, Colorado, wants geospatial data about fire roads from neighboring Pike National Forest (part of the USDA Forest Service (FS)) and El Paso County, Colorado, to use in wild fire mitigation, planning, and response. For some organizations, such as the U.S. Army and Air National Guard, such sharing with state and local governments is critical to their mission.

Other DoD organizations also need other government agency data and even industry data, such as utility company data, because OSD conducts Critical Infrastructure Vulnerability Assessments at installations that need utility infrastructure data.

Besides sharing with different parts of the U.S. government and the commercial sector, DoD organizations need to share with universities, NGOs, and allied governments. As already discussed, many installations already share geospatial data with scientists, and researchers from universities and environmental NGOs are conducting environmental or cultural resource research at their installations.

In fact, nonmilitary I&E geospatial data asset sharing is becoming more important to military installations and other parts of DoD for a variety of reasons. First, DoD is outsourcing and creating more and more public-private partnerships for diverse installation functions, such as installation housing and utilities, and needs to share geospatial data to help use, manage, protect, and maintain such facilities. Second, most military installations are no longer the isolated communities they were 10–20 years ago; they are both physically and politically closer to local communities. Communities expect more information and community stewardship from local installations, when dealing with such concerns as noise and environmental management. Third, the need for collaboration in homeland security and natural disaster response has increased. Last, the advancements in information and geospatial technologies, such as world wide web mapping applications, make cross-agency collaboration easier and more productive, so more agencies want to incorporate these technologies in missions, such as emergency response and natural resource management.

Many Barriers Exist to Successful Sharing of I&E Geospatial Data Assets

Despite the many examples of sharing I&E geospatial data assets, many barriers still exist to successful sharing. By successful, we mean sharing data to add value to mission functions, such as providing efficiency and effectiveness benefits. The ability to address and overcome these barriers will determine the future growth of successful I&E geospatial data sharing. The main barriers identified in our study, mostly through the interviews, are grouped into eight categories:

- security concerns and other data restrictions
- different IT systems, firewalls, and policies
- lack of communication/collaboration among different functional organizations and disciplines
- lack of knowledge about, interest in, or expertise with I&E geospatial data assets
- lack of data-sharing policy, standards, and contractual agreements
- unwillingness of data stewards, who want to control access to their data
- lack of on-going high-level program support and investments
- risks from sharing undocumented, poor-quality, and out-of-date data.

Each of these, not in priority order except for the first barrier, are discussed below. Similar categories are grouped together. In discussing these barriers, we briefly mention some of the steps that need to be taken to surmount such barriers that were suggested by our interviews and analysis. These suggested solutions foreshadow some of the recommendations made in Chapter Eight, where recommendations for the DISDI Office are presented.

Security Concerns and Other Data-Sharing Restrictions

The main barrier to I&E geospatial data asset sharing mentioned during our interviews was security concerns. The concern is that the information could be used by terrorists, criminals, or others who want to attack military installations or activities or engage in other destructive acts. Many basic I&E geospatial data assets, such as GIS boundaries, roads, and satellite imagery datasets, do not pose a security risk, because the information contained within them is so general and common knowledge, is available from other sources, or is not particularly useful for a potential attacker to use in actually planning an attack against an installation.[2] Some I&E geospatial data have sensitivities that evoke valid security concerns, making them difficult to share, such as data that provide information about security measures and practices. In other cases, potential risks are unknown.

Often, security staff want to restrict access to I&E geospatial data assets because there is no clear policy or guidance on identifying which geospatial information is sensitive. Because of such concerns, restrictions may be placed on some assets that do not enhance security. This restricts the benefits that could be accrued from more widely sharing these data. For example, APG GIO staff had planned to develop a web-based portal for sharing geospatial data assets across the installation, but security staff put a stop to the plan because of concerns about sharing this information across the web. Policy and procedures need to be developed to identify which data are sensitive and how such sensitive data can be shared so that the full benefits from data sharing can be realized. Addressing this security issue is especially important for web-based sharing and future data sharing in mission areas that involve collaboration with organizations outside the federal government, such as state and local governments involved in emergency response.

---

[2] For more information about how to identify geospatial data that pose a security risk, see Baker et al. (2004).

Another barrier related to security is that data cannot be widely shared because they are proprietary or have licensing restrictions. Such restrictions often occur when DoD purchases satellite imagery data of installations, and contracting agreements with the commercial vendors do not allow sharing outside the federal government. This can limit the sharing of such data for missions that involve nonfederal organizations, such as state and local governments involved in emergency response. However, during Katrina, commercial satellite companies allowed some of their imagery to be shared freely. Another barrier occurs when installations want to acquire and share private sector geospatial data about their installations, especially if they try to share data with more than a few key people at the installation. Since so many installations have outsourced infrastructure support, such as utilities, private companies develop, use, and maintain large amounts of geospatial data and may have proprietary concerns about them. Often, they will share data with installations but require that the data be treated as proprietary and strict restrictions are placed on who can access and use it.

Different IT Systems, Firewalls, and Policies

Another major barrier to sharing is that different military organizations have different IT systems, firewalls, and policies. Installation staff can have problems sharing data off base, because only internal staff have access to a geospatial system as a result of the base firewall. In fact, sometimes there are different firewalls even within the same building at an installation. In some cases, IT staff require lengthy approval processes before GIS software or applications can be loaded onto installation machines, do not allow geospatial applications to link into other installation databases, do not allow geospatial data sharing through the web, or do not allow data or software downloads from the web. Many of these IT policies and procedures have developed because of IT security concerns, especially those related to computer viruses. All these restrictions can limit geospatial data sharing.

This barrier can be caused by IT staff's lack of knowledge about geospatial information, management, and analysis systems and the benefits of sharing such data. Developing appropriate security procedures and processes for sharing geospatial information, and educating IT staff and management about the benefits, would help break down this barrier. Some interviewees also stated that official OSD policy about the need for and importance of sharing I&E geospatial data assets was needed to help them work with IT staff in addressing these issues.

Lack of Knowledge About, Interest in, or Expertise in Using I&E Geospatial Data Assets

Other barriers to sharing I&E geospatial data assets are the lack of knowledge of their existence and the technical expertise to take advantage of them. This latter barrier was found mostly with end users, especially at headquarters organizations, among more senior decisionmakers, and with OSD decisionmakers and analytical support staff. Such organizations, which can often benefit significantly from geospatial analysis and support, often lack the knowledge about how geospatial data assets and asset sharing can help their missions. If they are aware of the assets, they may lack the technical expertise to share and use them. For example, personnel at one OSD office had fundamental geospatial skills and access to strategic installation data from the DISDI Office, such as one-meter imagery of installations; however, since they could not store or process the 80 gigabytes of imagery data that they had acquired from the DISDI staff, they were using Google Earth in an analysis instead of the official OSD imagery. Another organization was using U.S. military installation boundaries from an Environmental Systems Research Institute (ESRI) federal lands database, even though the data were not very good, since they were easy to use and in the public domain. Furthermore, the analysts did not know how to acquire official OSD boundary data.

Interviewees stated that sometimes potential users do not want to learn how to use the new technologies, such as a GIS software program or even a geospatial web-based system. Many people resist trying new approaches, preferring to do things in an old familiar way, especially if the new approaches involve unfamiliar computer technologies and applications.

More education, dissemination, and technical assistance about I&E geospatial data assets, their availability, their advantages, and their use will help address this barrier.

Lack of Communication/Collaboration Among Different Functional Organizations and Disciplines

Another barrier to sharing I&E geospatial data assets is functional stovepipes. Namely, different functional organizations, such as public works, training, and environmental staff at an installation, do not communicate or coordinate much with each other, even though they all have developed and are using I&E geospatial data assets. At some large installations, different functional staff are spread across the base, which limits communication and geospatial data sharing. In many cases, such organizations do not publish their data assets so that others may see what is already available. The different organizations' terminology, interests, and software systems and approaches can also limit communication and data sharing. A common example is that many facility engineers are more comfortable using CADD systems for storing and managing information about their facilities, whereas environmental staff members often use GIS. At many installations, translation programs between CADD building databases and GIS databases are needed so that the systems can interface smoothly, especially since some of these data, such as the CADD building information, are being updated fairly frequently. Staff members may not want to take the time to learn how to do this or to invest resources in it.

In addition, geospatial data layers and tools are developed for individual functional interests and staff members may not take the time to see the synergies and commonalities with other geospatial data applications. Part of this reluctance to collaborate comes from not wanting to change the way things have traditionally been done. In addition, the cultural and political issues of the different organizations also contribute to this lack of sharing.

Management's supporting, investing in, and taking the lead in promoting the benefits of sharing data across functional organizations will help address this barrier, such as requiring coordination, sharing, and centralized management of I&E geospatial assets. Again, interviewees voiced the need for high-level OSD policy guidance stating the benefits of I&E geospatial data asset sharing to help address functional stovepipe challenges.

Unwillingness of Data Stewards, Who Want to Control Access to Their Data

The people who develop and maintain geospatial datasets often want to control who has access to them. Some data developers feel a strong sense of ownership and are reluctant to share data, especially with people they do not know. Often, people are willing to participate and share data and do so on a one-to-one basis, if they know the other person using the data and how the data will be used. However, others fear the wide-open data-sharing relationship when data are on a central network, because they do not know who is using their data or how. In such cases, they are reluctant to share their data outside the installation and do not want them to be available in web-based systems. They fear that other people will not properly understand the data and may use them inappropriately. In some cases, this is a valid concern, as will be discussed below. However, with proper metadata that describe the data and restrictions on its use, such risks can be minimized. For example, the metadata may state that the data are not of sufficient quality to be used in a court of law. An example is provided by the New York State Adirondack Park Agency when sharing its data with other organizations through the New York State GIS Clearinghouse. The description for selected GIS datasets states, "Data set should not be used for legal jurisdictional determinations."[3] Concerns such as these can be addressed by setting up proper procedures about how the data are shared and used. In some cases, additional information can be attached to the data; in others, data stewards may have more say about who has access to selected sensitive data. Having well-designed and properly implemented QA/QC procedures, as in the IVT process, can also help address such concerns.

Lack of Data-Sharing Policy, Standards, and Contractual Agreements

A major problem for sharing I&E geospatial data assets is not having sufficient standards development and implementation, policies, and contractual agreements about when and how to share them.

---

[3] See New York State Adirondack Park Agency (n.d.).

Standards are critical to sharing I&E geospatial data assets. Historically, dating back to the 1970s and 1980s, information system standards were classified into four main categories: data, processes, organizations, and technology. A range of standards exists within each of these standard classes. Data standards pertain to data directly, such as data content, classification, transfer, presentation, and usability. Process standards describe how to do things, such as methodologies to apply and procedures to follow. Data-processing standards include data collection, storage procedures, data-analyzing procedures, data integration, and QC and QA process standards. Organizations standards are the rules for assigning responsibilities and authorities to people who use the technology, such as rules about operator skills, tasks to be performed, and what data they need. They also include standards for human communication across organizations. Technology standards are the software, hardware, and system protocols that facilitate interface and interoperability among different technology systems.

This classification of standards is still useful today when looking at the standards needed to share I&E geospatial data assets. Many of these needed standards are developed by other organizations because of broader industry and U.S. federal government IT and geospatial technology standards. For example, the IT industry has developed standards to facilitate the sharing of data across the web, and the FGDC has developed U.S. federal government standards for data sharing, such as the FGDC Content Standard for Digital Geospatial Metadata. All federal agencies, including DoD, are required to follow the FGDC geospatial metadata standard in their development and acquisition of geospatial data. However, some standards are unique to DoD and I&E geospatial data assets. Such standards have to do with data, such as content and classification standards; process, such as data integration and QA/QC process standards; and organizations. For example, under the lead of the USACE CADD/GIS Center, the SDSFIE[4] have focused

---

[4] SDSFIE provide a standardized grouping (consisting of both graphic, i.e., symbology, colors, and linestyles, and nongraphic, i.e., database schema standards) of geographically referenced features, i.e., real-world features or objects depicted graphically on a map at their real-world locations. Each geospatial feature has an "attached" attribute table containing pertinent data about the geospatial feature.

on the development of graphic and nongraphic standards for GIS implementations at Air Force, Army, Navy, and Marine Corps installations, and for U.S. Army Corps of Engineers Civil Works activities.[5] The Services require that installations use SDSFIE, but many still do not because of the extra time needed to do so. Another problem is that different organizations use different standard versions, such as different parts of an Army installation using different versions of SDSFIE.

Standards about data content, format, and structure need to be developed and adopted to facilitate I&E geospatial data sharing. Lack of standardization of such items is a barrier to sharing within an individual Service and across DoD. First, different functional organizations name things differently; when identifying a building, some organizations use a building number, others a building ID, and others an asset number. Such differences are even greater across Services. A translation program must be built to share such data, which slows down interoperability and sharing. Second, different DoD organizations use different data formats and structures. The development and adoption of common data content, data dictionary, and data model standards are needed to address such issues. In addition, data-sharing models need to be developed. Third, organizations must invest the effort and money to follow standards, such as the metadata standard. We found that many DoD organizations still do not necessarily take the time to develop FGDC-compliant metadata or to meet SDSFIE. Creating metadata can be time-consuming and expensive so not everyone invests the effort, especially if they believe that they will be the only users of the data. Without the continued development and implementation of such standards, I&E geospatial data assets cannot be as successfully or broadly shared or discussed.

Currently, there is no official OSD policy about where, when, how, which, and with whom I&E geospatial data assets can be shared. Practices and policies are inconsistent across different installations, Services, and other DoD organizations. This lack of OSD policy guidance has limited asset sharing, because some organizations are reluctant to

---

[5] This U.S. national standard has been endorsed through the American National Standards Institute (ANSI) process and is also used outside DoD.

invest the time in developing the procedures, special agreements, and management approvals needed to share the assets. The lack of contractual agreements for data sharing, especially with respect to reciprocity, also limits I&E geospatial data asset sharing. Memoranda of understanding (MOUs) and agreements need to be developed but these often take time and a large amount of upper management scrutiny, especially when it comes to sharing data with individuals and organizations outside DoD. Organizations outside DoD, such as commercial firms and even local governments, may have licensing restrictions and proprietary concerns that can limit the ability to develop such agreements. OSD and NGA could help the situation by developing some standard MOUs for data sharing. The work that DISDI staff members are doing with the Project Homeland Pilot and the SERPPAS initiative represents good experiments for developing some standard agreements by DISDI in partnership with NGA.

Lack of On-Going High-Level Program Support and Investments

Another barrier is that organizations do not receive on-going long-term support from management for investments in geospatial data assets, technologies, and data sharing. We found this lack of support at all levels—installation, functional and Major Commands, Service headquarters, and within OSD.

At many installations, there is no funding for a dedicated geospatial staff. Geospatial data asset development is another responsibility of public works, IT, or environmental staff, but the main funding for those agencies is for mission-specific operations, which often take priority. Geospatial data assets are developed to support specific mission functions, and often data stewards do not have the time to invest in creating their data in the right standards, such as in developing metadata, which is important for sharing such data.

Some installations do not have the resources to support a dedicated geospatial specialist to develop, maintain, and use I&E geospatial data assets. Other installations, such as Camp Lejeune, Langley AFB, and APG, have been able to develop and maintain support for geospatial programs because the installation commanders and other management staff have seen and understand the benefits. All Service headquarters geospatial data asset staff are working to address such issues, but some installations and Services have made more progress than others.

Similarly, at the Major Command and functional levels, investments in I&E geospatial data assets are inconsistent. In some, where the benefits of using and sharing geospatial data assets have been demonstrated and are understood by management, funding and support have been more consistent. The U.S. Army's ITAM program and USAF Europe (USAFE) are two examples of organizations with strong geospatial programs and ongoing support. But in other cases, as was also seen at the installation level, organizations are investing in and supporting only specific projects or functional applications or geospatial tools; funding does not include the fundamental, consistent investment in basic I&E geospatial data and capabilities.

The Service headquarters organizations, i.e., the GIOs, have all been created in the last few years and they, especially the Army and Navy, could benefit from more management support and investment to develop their offices and Service-wide programs to ensure consistent long-term investments in I&E geospatial data sharing.

Similarly, many OSD offices do not have the support to invest in I&E geospatial data asset development, use, and sharing. In some cases, they do not need to if they can secure the technical assistance and help of DISDI. However, DISDI itself is very new and will need to maintain and even expand its management support and resources to provide more assistance and support to other OSD offices. For some OSD offices, such as the environmental ones, there may be enough need and benefit from geospatial assets over the long term, given current trends, that OSD environmental staff should invest more directly in I&E geospatial data assets rather than relying on the DISDI office.

Risks from Sharing Undocumented, Poor-Quality, and Out-of-Date Data

Another barrier to successful I&E data sharing is that data could be shared that do not add value to the mission function because they are undocumented, poor, or out-of-date. In some cases, poor-quality or out-of-date data could actually hurt a mission by providing inaccurate data. An example occurred at an Army installation in the Southeast where the GIS dataset of surface danger zones (SDZs) had been shared with a local county government because of safety concerns. This SDZ showed that part of the exclusion area identified to protect people from live firing during training extended off base. However, the SDZ dataset was out-of-date and incorrectly showed a smaller SDZ area. Using this out-of-date dataset, the county was inadvertently about to allow residential development within the SDZ area near the base. Fortunately, the installation provided the county with a current SDZ data file and development was prevented.

Similarly, missions that involve timely responses, such as emergency response and homeland defense, can experience significant problems when poor-quality geospatial data are used. For example, when dispatching an emergency medical vehicle, having an accurate street address and road network database can be a matter of life and death. Unfortunately, geospatial datasets are sometime assumed to be accurate when they really are not. GIS and image data might be assumed to be always accurate, but significant data-quality issues can exist, especially for image data that require more interpretation and processing.

If data do not have metadata, the user cannot determine data quality and there is a risk that the data will be used for unintended or inappropriate purposes.

Evolving I&E Geospatial Data Asset Applications and Use

Given the current use of technologies and how the technologies are evolving themselves, there are some implications about how the use of I&E geospatial data assets will likely evolve in the future. The four key trends discussed below outline how the use of I&E geospatial data assets will likely evolve in the future:

- increased use of web-based spatial portal systems
- increased use of real-time information
- more centralized and enterprise approaches

- more integration and sharing of more detailed information from diverse sources.

However, these trends will be more limited if the barriers just discussed are not addressed and overcome. For example, given the distributed nature of sharing in the future, successful use of such applications will depend even more on the use of standards and good QA/QC of the data. Similarly, security concerns and IT firewall issues will also have to be addressed.

Increased Use of Web-Based Spatial Portal Systems

In the future, if the security and firewall issues just discussed can be addressed, technology and current application trends indicate that there will likely be more web-based portal systems, making more I&E geospatial data and sophisticated applications available to more users. As discussed above, spatial portals are web-based gateways through which users can disseminate, discover, and access geospatial information. A successful portal connects desktop users with disparate data holdings and applications through the web. Spatial portals, both within individual installations and for regional, national, and international data-sharing initiatives, provide a way to order, manage, discover, and access distributed geospatial services. Technology advances have made portals an effective and efficient way to distribute and share geospatial information.

To understand likely future trends for portals, it is useful to discuss the different types of portals, because some make the data available to a much broader user community, not just geospatial technology experts or analysts. There are three main types of portals: catalog, application, and enterprise. Catalog portals maintain indexes or catalogs of available geospatial data and information services. A catalog portal provides indexed geospatial metadata that users access and search to view and find maps and geospatial datasets of interest. For example, DISDI provides the DISDI Metadata Portal, where users can search and discover the availability of I&E geospatial data by viewing detailed metadata describing those data resources and access geospatial data holdings present in the Services' I&E geospatial program portals and repositories. Such catalogs usually provide information to GIS professionals and are not as effective for distributing geospatial information to less-knowledgeable individuals. Catalog portals most likely will be used when an organization wants to provide information about the location of I&E geospatial data assets without providing the actual assets themselves. Such systems are especially useful when strong access control is desired, such as when security, organizational, or licensing restrictions hinder freely sharing data.

An application portal is one where the user can perform some sort of geospatial application through the web. Application portals are usually for a well-defined audience or those with specific application requirements. Application portals provide web-based mapping tools to allow users to view and work with the data they find (for example, geo-processing tools such as route and distance finding, geo-coding, and buffering). The distinction between application and catalog portals is blurring as catalog portals increasingly add visualization and analysis tools. An example of an application portal can be found at the IMA KORO, which has developed a GIS Repository for the region that users from Army installations throughout Korean bases can access through the web to obtain GIS data and tools. The system provides a standard toolset for planning, engineering, and facility management, including the web-siting tool, web evacuation permit system, and web planning tool. Korea headquarters IMA staff, customers, and individual camps throughout Korea use the system. Korea Wide Area Network (KWAN) access and AKO authentication are required for access. The use of such application portals for I&E geospatial data assets will likely continue to grow, because they help meet the needs of different missions and can often be easily used by functional staff who have little knowledge of GIS or other geospatial tools, as seen with both the Langley AFB floodmap tool for emergency response and the USAREUR web-based ITAM Mapper for viewing training areas.

An enterprise spatial portal integrates geospatial data and functionality with business enterprise systems across that enterprise. Only in the last few years have enterprise systems been developed for the management, use, and sharing of geospatial information. For example, Fort Rucker has an Enterprise Geographic Information System (E-GIS) for managing and sharing GIS information. It provides support to senior leaders, garrison staff, unit commanders, and trainers. Through the web, diverse users from across the installation access, use, and update geospatial data to help with training and installation management. Organizations such as range operations, force protection, airspace management, environmental, natural resources, engineering, and the plans, analysis, and integration office currently operate within the enterprise structure. Some enterprise spatial portals integrate geospatial data and functionality with business enterprise systems for a few systems; others integrate many, as can be seen in the Camp Lejeune example. Enterprise portals require much more investment to develop and maintain the system and link it into other functional databases. However, given the benefits of having a centralized source for maintaining, updating, and accessing data, more and more organizations are developing them, such as Langley AFB's and Ramstein AB's viewers.

Many geospatial portals are combinations of catalog and application portals, application and enterprise portals, or all three types. For example, the four installation examples mentioned above, Camp Lejeune, Ramstein AB, Langley AFB, and Fort Rucker, are all combinations of application and enterprise portals. The development and use of such combined application and enterprise portal systems for I&E geospatial data assets are likely to increase. These application and enterprise portals enable I&E geospatial data assets to be used much more widely, including by individuals who know nothing about GIS or other geospatial technologies. However, the barriers just discussed will need to be addressed before the development and use of such systems grow significantly and enable I&E geospatial data assets to reach the broadest set of mission users.

Increased Use of Real-Time Information

Another important trend with current technology that was not feasible until recently is the use of real-time or close-to-real-time information in real-time geospatial applications. This has come about because of the continued integration of information technology and monitoring systems, field technologies, and the distribution of real-time information through the web. These live applications use real-time information from the field, such as weather, traffic, environmental monitoring data, and other current conditions. In the previous chapters and the appendix, several examples are given where geospatial applications accessed real-time data distributed across the web, such as APG and Ramstein AB GIS staff putting current incident information into the installation GIS system for use by the Emergency Operations Center (EOC) during emergency response situations; the USAF accessing the National Oceanic and Atmospheric Administration's (NOAA's) latest hurricane tracking when preparing for Hurricane Rita; and IRRIS incorporating real-time weather and traffic data as it tracks and manages military logistics.

Integrated GPS systems, on-vehicle location systems, laptop computers, and hand-held field technologies can all be used to integrate real-time information from the field about storm damage or other environmental conditions, such as during a disaster response and recovery, into geospatial systems. For example, in May 2004, 52 federal, state, and local agencies and other organizations participated in Global Mirror, a DHS and Federal Emergency Management Agency (FEMA) emergency preparedness exercise in Colorado Springs, Colorado. The exercise used GIS during a scenario involving a weapons of mass destruction (WMD) terrorist incident. The exercise also crossed the government jurisdictional and geographic boundaries of Peterson Air Force Base, the city of Colorado Springs, and El Paso County, Colorado. In this exercise, emergency responder vehicle locations were tracked almost in real time using advanced vehicle location technology.

Ongoing monitoring systems, of weather, air quality, and water quality, for example, can be linked into I&E geospatial data assets to help in planning, management, assessment, and operations for environmental, emergency response, homeland security, force protection, medical, training, and other missions that require current data. For example, at Fort Benning, geospatial information about the location of weather stations and real-time weather data is used to help with controlled burns and for scheduling training exercises that depend on current weather conditions. Such applications will likely increase in the future as such monitoring systems are integrated into I&E geospatial data applications. However, these applications depend on being able to integrate and access quality information in close to real time, which means that the barriers that affect interconnectivity, such as IT firewalls and standards adoption, and the barriers to having high-quality data readily available need to be addressed before these applications can spread more widely within DoD.

More Centralized and Enterprise Approaches

The benefits of having centralized and enterprise data approaches, where many people both within and outside organizations can access and use data, will likely encourage more such approaches. Centralized approaches mean that there are centralized sources of geospatial data or common points of access for distributed data using standards to obtain interoperability and data sharing. Such approaches are already spreading, within both organizations and applications. Both the Navy and the USMC are using centralized approaches for many of their geospatial data asset applications, such as the Navy's RSIMS, a Navy-wide Internet map viewer for Navy installation data, i.e., a global GIS system for all installations. Similarly, at the installation level, some of the greatest benefits have accrued from centralized and enterprise approaches that centralize most of the data management and share the data broadly with diverse users, such as Langley's basewide "Langley GeoBase MapViewer," Keesler AFB's GeoBase system, Aberdeen's centralized GIS system, and Fort Sam Houston's Enterprise GIS, an integrated installation-level GIS that gives users broad access to geospatial information. At the headquarters level, all the Services except the USAF are creating centralized data repositories, such as the Army GISR and the U.S. Navy GeoReadiness Repository. The USAF's Major Commands are taking a centralized approach to sharing data by having Major Command Geobase web portals containing installation CIP data and some mission data assets. By spring 2006, AMC, USAFE, and ACC had taken such approaches and were sharing this information more widely through the USAF portal. At the OSD level, the DISDI Office has developed the DISDI Portal for centralized OSD I&E geospatial data assets. However, as with the other future trends, such growth and widespread use of such approaches depend on addressing the barriers discussed above, such as implementing standards, having quality data, and addressing security and IT issues.

More Integration and Sharing of More Detailed Information from Diverse Sources

Closely related to the enterprise approaches is the likely future where more detailed information from diverse systems will be integrated and shared within geospatial systems. Such information includes more detailed geospatial data as well as other functional information from diverse installation and other military database management systems. First, geospatial information is more widely available at a finer and finer scale, accurately placing elements of interest within centimeters or less. For example, CADD systems provide detailed accurate internal data about buildings and other assets. Such geospatial information includes internal geospatial building information; the location of equipment, furniture, and key infrastructure, such as ducts, pipes, wiring, and even sprinklers; as well as more detailed information external to buildings, such as individual tree and plant species locations and road and parking lot pavement conditions. Such data are already being integrated into some I&E geospatial data applications, as is illustrated in a number of examples in the appendix—from Marine Corps Air Station (MCAS) Cherry Point using building sprinkler information to the USAF tracking individual tree species on parts of installations. The evolution of Building Information Model (BIM) approaches will see more detailed building information integrated with other geospatial information in future geospatial applications. BIM is a three-dimensional computer model that serves as a single repository for the drawing and database information traditionally associated with the design, construction, and maintenance of a building, including detailed building plans and maintenance requirements. As this concept develops, more detailed internal building information will be used and shared. However, security issues must be addressed as more detailed information is used and shared.

Second, it is easier to integrate and bring in data from other military database management systems within geospatial systems, and these systems are tracking more and more information about diverse resources. The appendix and Chapter Three provide numerous examples, where the geospatial systems linked into other functional databases with information about such resources as manpower, equipment, and medical capabilities. For example, NAS Patuxent River combined building floor plan details with janitorial services and tenant information from other installation databases in their GIS analysis to more efficiently allocate janitorial costs. Again, barriers regarding system interoperability need to be addressed for such future uses to spread across DoD.

In conclusion, I&E geospatial data asset use and sharing are likely to continue to grow. However, how widespread this growth is and how fully some of the technology trends affect this growth depend on addressing some critical barriers that limit sharing. In Chapter Eight, we present some recommendations for the DISDI Office to address such barriers and help enable more widespread use and sharing of I&E geospatial data assets across DoD. Next, we discuss how to assess the mission effects of using and sharing I&E geospatial data assets.

CHAPTER SEVEN

Assessing the Mission Effects of Using Shared I&E Geospatial Data Assets

In Chapters Three and Four, we examined the diverse mission areas that can be supported by I&E geospatial data assets. We noted that organizations and individuals experience a wide range of mission effects from using and sharing geospatial data assets. Some are common across mission areas, geographical areas, organizations, and echelons within DoD. Common benefits include time savings, cost savings, cost avoidance, improved situational awareness, improved communications, and better decisionmaking. Other effects are unique to specific installations or missions.

In this chapter, we examine how to assess the mission effects of using and sharing I&E geospatial data assets. First, we describe the various types of mission effects and then we present our methodology for assessing them. Finally, we apply the methodology to some examples.

The Diverse Effects from Using I&E Geospatial Data Assets

We found that the use and sharing of I&E geospatial data assets have many different types of mission effects. As we will show, they are seen at all levels within the Department of Defense—from an individual office on an installation to the Office of the Secretary of Defense. Our definition of effects is broad and includes the attainment of desired outcomes by the individual organization developing, using, or sharing the assets and any other outcomes experienced by any organization from that asset development, use, and sharing. Namely, effects from the use of I&E geospatial data assets are not just for the organizations using the assets; they touch other organizations both across and outside DoD. We have found that using and sharing geospatial data assets generates four categories of effects:

- changes in efficiency
- changes in effectiveness
- process changes
- other mission effects.

At least implicitly, we are suggesting that the goal of use and sharing is to improve the efficiency and effectiveness of organizations' efforts to attain mission objectives, although in some instances, it may have an even more direct and immediate bearing on mission attainment.

Other researchers studying the effects of using and sharing geospatial data assets have offered different taxonomies, and no seminal or gold standard list of benefits exists; authors use what is most appropriate for their subject. Most have focused on efficiency benefits, because they are easier to quantify and to subject to benefit-cost analysis studies. For example, in an analysis of 62 U.S. federal government case studies of applying GIS during the early 1990s, benefit-cost ratios for efficiency benefits ranged from 1.2 to 5.6.[1] The state of Montana has looked at both efficiency and effectiveness benefits within state and local governments throughout their state to improve the use of geospatial information. However, the state had a difficult time quantifying effectiveness benefits.[2] Examples of benefits that are more difficult to quantify include improving decisionmaking, improving information and services provided to customers or the general public, increasing public safety, and improving environmental quality or other quality-of-life features. These examples all illustrate how difficult it can be to quantify the benefits of using and sharing geospatial data assets and, if they are quantified, how easy it is to underestimate the full benefits,

---

[1]   Gillespie (1997). See also Gillespie (2000, 1994a, 1994b).

[2]   McInnis and Blundell (1998).

which is why we take a different approach here and try to look at the full range of effects. One distinction we have made in our research that is particularly intriguing, and quite relevant to this study, is between anticipated and unanticipated effects.

Organizations that invest in the collection and application of geospatial data assets most often seem to anticipate efficiency and effectiveness benefits to accrue from their investment. Using digitized geospatial data saves time in verifying the location of a site and eliminating the need to visit it, allows for faster information processing, and so forth. However, organizations often underestimate the extent of those gains. For example, once the data and related systems are in place, organizations often identify additional uses that further improve efficiency, or they find that the intended use of the assets generates benefits that were never anticipated, such as improved communications between two offices. This is an example of an unanticipated benefit that comes from the use and sharing of I&E geospatial data assets.

Changes in Efficiency

Efficiency effects relate to the amount and type of resources that organizations use to create products or perform a mission. With an efficiency effect, the type of output or ultimate outcome is the same but either the input is different or the amount of output or the outcome is different. For example, the use of I&E geospatial data assets can save tremendous amounts of time, affecting how many people work within an organization and how they spend their time. As a result, organizations could reduce manpower (and therefore costs) or increase output. With resource inputs, we must also consider the cost of investment and maintenance of geospatial data assets. These assets can clearly save time and money, but they are not free. We will address investment and maintenance as well as efficiency gains in this section.

Efficiency effects are grouped into two categories:

- effects on time and manpower
- effects on other costs.

Effects on Time and Manpower. Many uses of I&E geospatial data assets have some sort of effect on time savings or manpower. Using geospatial resources in a GIS or web-based application saves time in processing and analyzing data and in producing and processing maps, reports, and PowerPoint charts. Often, savings are significant compared to doing these functions by hand without the computerized geospatial information. For example, consider the geospatially web-based Dig Permit Tool at Keesler AFB, Mississippi (see the discussion in the appendix). This Dig Permit Tool allows a dig permit request to be processed and approved in about four hours and, under ideal circumstances, within an hour if each reviewer sees the request and processes it immediately. Previously, a request had to be hand-carried to eight offices around the base, and the request took at least a full day to process. Since Hurricane Katrina, about five or six dig permit requests are submitted per day, whereas before (or during "business as usual"), about two or three requests were submitted per week. Thus, by being able to automatically route the requests to the appropriate approval authorities, not only is the amount of processing time cut in half for a single request but a huge backlog in construction projects is avoided.

An important manpower savings also occurs from the sharing of I&E geospatial data assets. Acquiring, maintaining, and updating geospatial datasets is not cheap. In the 1990s, data collection accounted for about 60 percent to –80 percent of the costs of a fully operational GIS system.[3] It still is a major cost of I&E geospatial data asset development and much of this cost relates to the manpower required to process, maintain, and update the data. In some cases, the cost is for a data purchase, when some other organization has incurred the costs of creating/acquiring the data, such as buying processed imagery data from an aerial remote sensing firm. When the geospatial data are shared, organizations no longer have to invest the manpower and dollars to acquire and maintain their own versions of the same data. In fact, such savings was a main motivation for the implementation of the

---

[3] Purchase of the GIS constituted 10 percent to –30 percent of the costs, and the remaining amount was spent on such other items as training and administration. Bernhardsen (1999, p. 151).

GeoBase program within the USAF. Such savings are often called cost avoidance, because by sharing data, an organization avoids a cost. We discuss such savings in two subsections below, under other costs and again later, where we quantify some of the savings from the sharing of installation imagery.

Sharing geospatial products through web-based systems with many users also can create significant manpower savings. By providing users with a web system to analyze or map geospatial information themselves has saved GIS shops considerable time in conducting analyses and producing maps. For example, the USAREUR ITAM program's ITAM Mapper web system has had significant use and has saved the USAREUR ITAM program staff significant amounts of time in not having to produce as many maps. Between January 2003 and June 2005, there were over 3.1 million hits at this web site with 530,000 page views, 17,000 visits, and 6,100 unique visitors, with an average time per visit of 14 minutes.

It is important to note that such systems have an upfront investment time in which to develop the geospatial data and application, but it is offset by the time savings in using the application, such as with the USAREUR ITAM Mapper example. Similarly, in the dig permitting example, the basic geospatial data were already developed, so only time and manpower were needed to develop the tool, which was a minor cost compared to the savings produced by its use.

In some cases, time is saved at multiple organizations and not just within DoD. For example, the communication guide for NPS managers that the Air Force and NPS jointly developed eliminated time spent in both organizations when responding to noise complaints from visitors to national parks. For each noise complaint, USAF headquarters staff used to spend a few hours investigating it, staff at the installation spent a few hours, then USAF headquarters staff had to write a memo, which went to the Secretary of Defense, who sent it back to the Secretary of the Interior. Similar staff time was spent in the NPS chain of command. Now the NPS and local Air Force base personnel can deal with the complaints at the local level, which saves staff time at both organizations' local and headquarters organizations.

Perhaps the most detailed study of time savings was conducted by the Patuxent River Naval Air Station, which published a thorough return on investment (ROI) study of its IT investment program in 2000. Data were collected for the study through interviews with staff, and routine work was timed using a stopwatch to measure precise time savings. Several of the processes that experienced time savings applied geospatial data assets. For example, the Public Works Department Environmental Support Group (ESG) must respond to 10–12 data calls per year from organizations such as DoD, the U.S. Fish and Wildlife Service, and the Maryland Department of Natural Resources regarding wetlands, archaeological programs, conservation expenditures, and a host of environmental issues. By using GIS to conduct acreage calculations, perform trend analyses, and generate maps, in tandem with developing a database and applications system known as the Environmental and Natural Resources Management System, ESG was able to process more data calls in less time. Table 7.1 shows the number of major data calls processed by ESG, and the amount of time spent responding to the calls, throughout the 1990s. According to the original study, the table "shows a reduction in effort through IT investment, despite an increase in the number of data calls required. . . . Small time savings realized on numerous minor calls and queries would increase these savings even more."[4]

Saving time on a task means that staff time can be dedicated to other tasks or staff positions can be eliminated altogether. In general,

Table 7.1
Time Savings by Patuxent River Public Works Department Environmental Support Group

Data Calls	1993–1996	1997	1998	1999
Number of major data calls per year	6–10	10	10	12
Average number of hours per data call	16	12	10	8
Total hours required	96–160	120	100	96

SOURCE: Patuxent River Naval Air Station.

---

[4] *Return on Investment Policies, Concepts, and Methods for Installation Life-Cycle Management* (2000, pp. 69–70).

time savings are easier to calculate than manpower effects, because the benefits are most often realized through greater output or a different focus of staff effort, not reduced headcount. However, although we attach a dollar value to time savings, we do not make an implicit assumption that time savings result in reduced manpower.

It is important to remember that the time and manpower effects are not all positive, however. Although an environmental office or a construction office might save time in processing permits or issuing reports, some amount of time must be spent in teaching staff how to use and maintain the data and applications. Moreover, as installations and other organizations invest in IT, the demands placed on the technical support personnel will grow.

Time and manpower effects obviously can touch organizational costs. In fact, most IT investment's return on investment or cost-benefit analyses focus on the very tangible time and manpower effects and then monetize them or calculate the amount of money that could be spent or saved as a result of the effects. As noted above, when there are time savings, the manpower effect is more likely to be seen in staff taking on additional responsibilities, not in staff being cut. Thus, the benefits may not be observable as "savings" in a budgeting sense; instead, they are seen in increased productivity, which we discuss in a subsequent section.

Effects on Other Costs. Besides monetized labor savings, other types of cost savings have been realized as a result of using and sharing geospatial data assets. An example of common cost savings was in using geospatial data assets to perform a job more accurately, saving money from performing the task poorly. Another example was using I&E geospatial resources to more accurately assess the location of sites for building new structures and other assets; this use saved dollars by reducing the cost of accidentally damaging utilities or making other construction errors. For example, the Department of Public Works at Aberdeen Proving Ground estimated that after GIS analyses were used in planning construction projects, the first-year cost savings in reduced damage to utilities near construction sites was about $115,000 (in 1992). Similarly, the Department of Public Works at Langley AFB uses a dig permit application to help assess the potential effects of dig locations on underground petroleum storage tanks; this also helps reduce costly accidents. Although most petroleum tanks at Langley AFB are out of service, it would cost between $12,000 and $15,000 to replace a damaged one.

A related effect is cost avoidance, as was briefly mentioned above. This often occurs through sharing of geospatial data assets, because a single organization can collect and manage the data assets and share it many times. A good example of cost avoidance is the acquisition of imagery, such as imagery cost sharing at Malmstrom AFB, Montana. Air Force Space Command (AFSPC) provided $200,00–$240,000 toward an estimated $2.4 million effort by the Farm Service Agency's (FSA) National Aerial Imagery Program (NAIP) to obtain orthorectified imagery across all of Montana. AFSPC's participation level was based on Malmstrom AFB's direct area of interest, covering 10 percent of Montana and the area of responsibility for emergency response and contingency operations, which encompasses all of Montana. Below, we discuss issues related to quantifying cost avoidance through the sharing of geospatial imagery.

Changes in Effectiveness

Effectiveness benefits result when I&E geospatial data assets are used to perform a task that would not or could not have been done without the assets or they improve the quality of the task being done. For discussion purposes, we group effectiveness effects into two broad categories:

- improved operations, decisionmaking, and planning
- performing a new task or providing a new service that could not or would not have been done without the assets.

Effectiveness effects are more challenging to measure or value than efficiency effects, since it is often difficult to attach dollar figures or other metrics to something as nebulous as the quality of decisionmaking. Still, those we interviewed often reported that having geospatial data and applications improved situational awareness, improved their decisionmaking, and increased their confidence in the decisions they made. A more tangible measure of effectiveness is an organization's ability to perform a new task or provide a new service, although these are also hard to quantify.

Improved Operations, Decisionmaking, and Planning. The sharing of geospatial data assets affects operations, decisionmaking, and planning at all levels of DoD. Many I&E geospatial data assets are being developed and shared to improve mission operations, analysis, planning, and decisionmaking. The specific improvements will usually differ depending on the mission types. We provide some examples below for the mission types discussed in Chapters Three and Four. For more examples, see the discussion in the appendix.

First, we discuss a common operational benefit from sharing I&E geospatial data assets—improving the quality of the geospatial data itself. Geospatial data sharing can improve data quality, because more individuals see the data and find and correct errors.[5] This benefit often occurs when other geospatial professionals and other experts, who are very knowledgeable about the characteristics that the data represent, view and use the data, because they are likely to notice any potential data problems. For example, when showing environmental data about habitat, vegetation, or species locations to different installation natural resource specialists, or when showing GIS utility data to utility maintenance crews, they may notice some data problems, because they know what is happening on the ground at the base. A specific example occurred when ASHS model experts applied the ASHS model to different USAF installations. Since ASHS requires very detailed and precise data, some of which often need to be gathered from the field, when using ASHS, these experts noticed some data problems and have cleaned up installation CIP data (see the discussion of ASHS in the appendix).

For base planning, operations, and management, I&E geospatial data assets help improve decisionmaking, management, operations, and the accuracy of assessments supporting such operations. Such improvements result in better placement and siting of new facilities; better infrastructure and facility construction, management, and oversight; and better use of construction and maintenance resources. For example,

---

[5] Sommers (1997, pp. 10–11).

the dig permitting applications discussed above improved the accuracy of locating and deciding where to dig. Another common example has to do with using I&E geospatial data assets to help plan the location of new buildings and other assets on an installation. By being able to view all the relevant information together in a GIS system, including the environmental constraints, installation staff can make better decisions about where to site things (see the example in the appendix). Improving contractor oversight, whether runway construction workers or grass moving contractors, was another facilities management effect.

Missions such as emergency response, homeland defense and security, safety, and antiterrorism/force protection all have benefited from improved preparation, planning, and response times, better and more coordinated decisionmaking, and better use of resources as a result of I&E geospatial data assets, which can help save lives and protect property. With the Project Homeland initiative (as discussed in Chapter Three), by having a common situational view of an incident, the commanders and first responders at Fort Carson, the USAF Academy, El Paso County, the state of Colorado, and DHS can plan, coordinate faster, and more effectively respond to any type of regional emergency incdent. Similarly, I&E geospatial data assets help improve the assessment of critical infrastructure vulnerabilities, which helps DoD organizations prepare better for a homeland security/defense incident. Operation flight safety has also improved, saving lives and avoiding the loss of expensive aircraft, such as by using I&E geospatial data assets to better manage wildlife near runways to reduce bird strike hazards (see the discussion in the appendix of BASH activities).

Use of I&E geospatial data assets in environmental applications helps improve environmental management and quality, such as improving the quality of forest, ecosystems, and the land, thus helping sustain both training and wildlife. For example, at Fort Bragg, North Carolina, over 480 miles of firebreaks were constructed during the 1970s without consideration for topography or the wetlands they crossed. After over 30 years of heavy use and the original disregard to their location, these firebreaks are now causing significant erosion problems throughout the installation. GPS field data about firebreak erosion problems were entered in a GIS and combined with other GIS data, such as natural resource and training range information, to help assess different mitigation options including identifying where highly eroded firebreaks could be closed and revegetated or rerouted, and where to enforce best management practices in their maintenance.[6] This geospatial analysis is helping to address the erosion problems to support training and to improve the quality of the land, ecosystems, and watersheds at Fort Bragg. Other effectiveness benefits include reduction in noise complaints and improving the preservation of cultural resources.

For military health missions, the use of I&E geospatial data assets has helped improve the planning, management, tracking, and assessment of military health assets and potential health threats. For example, the USAF Surgeon General Modernization Directorate has developed a GIS-based computer application, called Community Health and Medical Program (CHaMP), that provides integrated geospatial disease surveillance and outbreak detection, which can be used to improve the tracking of influenza outbreaks, such as the potential spread of avian flu (see the appendix). Another effectiveness effect is to improve the quality and quantity of medical services provided to our troops. For example, the OSD Health Affairs TMA/Health Programs Analysis and Evaluation Directorate uses its "Military Health System Atlas," and other I&E geospatial data assets, to help improve the medical services provided to U.S. military and other beneficiaries (see the appendix).

Similarly for the MWR mission, I&E geospatial data assets have helped improve the management, quality, and quantity of services provided to the military community. They help improve the management and placement of such facilities, whether a golf course, jogging trail, or military family housing. An example of improving service occurs with DoD schools in Europe, where GIS data and analyses were used as a classroom space management tool to better allocate and schedule fluctuating student populations and teachers to classes and classrooms (see the appendix for more details).

For strategic basing, I&E geospatial data assets helped improve the decisionmaking process in BRAC, as discussed in Chapter Five.

---

[6] For more details on this analysis process see Frank (2005).

For example, the OSD BRAC office requested and received about 350 installation maps to better understand the distribution and proximity of the many DoD installations being examined as part of the BRAC process. This helped BRAC decisionmakers understand the strategic, logistical, economic, and political implications of the recommendations being made for base closure and consolidation.

Effectiveness effects for training include facilitating more time on training ranges/testing facilities, improving the quality and safety of the ranges/facilities, helping keep the ranges/facilities open, and maintaining their operational flexibility. For example, as discussed in Chapter Three, by flying in the realistic Fort Hood simulation model (which uses I&E geospatial data) before going out on the training range, A-64 Apache helicopter pilots have cut the amount of time they need to spend on the gunnery range by about one-third. Many installations, such as Camp Lejeune and Fort Benning, combine range and environmental I&E geospatial data assets to more effectively manage T&ES concerns, such as the RCW, and facilitate more training time and space on the installations. Similarly, such training installations also use I&E geospatial data assets to improve the siting of a new training range or testing facility by minimizing the safety and environmental effects involved in locating, designing, and building it.

As discussed in Chapter Four, warfighting missions have also improved planning, operations, and decisionmaking, because of the use of I&E geospatial data assets. The assets have helped facilitate more rapid and effective logistics and deployment, such as helping with force beddown decisions for operations in Afghanistan and Iraq. The assets have also helped to improve base camp and forward operating site (FOS) management and operations, by using I&E geospatial applications, such as the RMTK and the CAPP. There has also been improved force protection and safety at such FOSs from the use of I&E geospatial data assets. The assets also help improve strategic planning, wargaming, and intelligence terrorism threat assessments.

Performing a New Task or Providing a New Service That Could Not Be Provided Before. Another type of effectiveness effect is the ability to perform a new task or provide a new service that could not or would not be done before. By integrating and sharing I&E geospatial data assets, new tasks become feasible that were not feasible before because of time, manpower, or budget constraints. For example, as just discussed, senior OSD leaders through the OSD BRAC office requested numerous installation maps over a two-month period, often expecting sets of JPEG map products within an hour. The DISDI Office was able to perform this task only because of the use of the IVT data.

New services are also provided as a result of the use of I&E geospatial data assets, as illustrated by a field mapping application for soldier training. The Army National Guard Training site at Camp Ripley, Minnesota, offered a new product to soldiers—a GIS-based kiosk in the range control office that allows soldiers to print their own maps of the training area. Using a touch screen on the kiosk, the user could zoom into the range or training area and could select from a limited set of GIS layers to include, such as the aerial imagery, buildings, training areas, and wetlands. Then the user could print out an 8-1/2 by 11 inch or C-size map. Figure 7.1 provides an image of this touch screen interface right before the user is about to print out a map.

Another kiosk at the billet office allowed soldiers just arriving to view maps of the base.[7] This new service provided timely in-the-field maps to soldiers—providing orientation and navigation to many who had never been to the base before. It also was used to help schedule training range time, since it linked into the range management database.[8]

Process Changes

Many applications of I&E geospatial data assets resulted in process changes, such as changing an analysis process, organizational process, or worker communications. Process effects are often an unanticipated benefit and seem less tangible than even effectiveness effects, but they are no less real and can be very important to individuals

---

[7] The kiosks were dummy terminals and the system was run through a server. There were some technical difficulties with the server going down frequently. The system was no longer in service in January 2006, but it is being redesigned and will be used again.

[8] It is interesting to note that this system was designed to reduce maps requests from geospatial staff, which it did; however, staff members found that they got new and more sophisticated requests as users learned about the GIS and wanted more on their maps.

Figure 7.1
Screen Image for Camp Ripley GIS-Based Kiosk

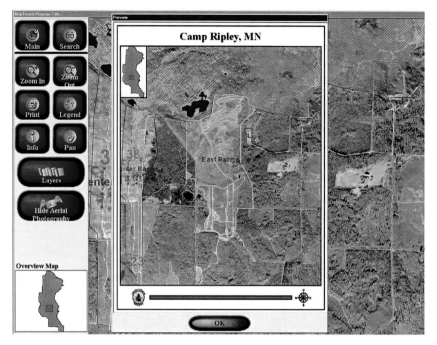

SOURCE: Image courtesy of the GIS Manager, Minnesota Army National Guard, 2006.
RAND *MG552-7.1*

and organizations that use and share geospatial data assets. In addition, they may result in efficiency and effectiveness gains. We have grouped process effects into two categories for discussion purposes:

- improved communications and working relationships
- other process changes.

Improved communications and working relationships were separated out because we found that they occurred in many application examples. This process effect helps address a key barrier to data sharing—the lack of communication/coordination between different functional organizations. Although we will discuss different types of process effects separately, they often go hand-in-hand, just as the various efficiency benefits do. Changes to an analytical process can easily lead to improved working relationships.

Improved Communications and Working Relationships. Geospatial data can quickly convey a large amount of information to people visually, and this can speed and facilitate communications between individuals and groups and improve working relationships. This was attested to by many of the people we interviewed. This benefit is a common second-order effect and can occur across diverse organizations and levels both within and outside DoD.

First, the use of the assets helps improve the communications and relationships across functional organizations, such as environmental, planning, installation management, and training staff. For example, at Travis AFB, a web visualization tool of installation geospatial data is used to see various land-use constraints for a proposed building project. The data reveal the presence of environmentally contaminated soil, sensitive plant species, or communication lines that might be disturbed by digging. Environmental staff, communication line staff, and civil engineer planners communicate more with the system, because they see everyone's data integrated in the GIS system and can see and discuss any potential problems. Similarly, an installation commander at a different installation found that by pulling up a map, everyone had a visual representation of what was being debated or discussed. A key benefit of this ability was improvement in the communication flow between groups that might be advocating different positions.

Second, communications are improved with headquarters organizations and other senior DoD decisionmakers. In the BRAC process, the maps created with IVT data by the JCSGs and the Service BRAC offices made it easier for these organizations to communicate their positions and decisions to general and flag officers and other BRAC JCSG and Service decisionmakers.

Third, using I&E geospatial data assets helps improve communications and working relationships with organizations outside DoD, as happened with the communication guide for NPS managers that the Air Force and NPS jointly developed. A common example of improving working relationships with local governments occurs in emergency response situations. For example, Vandenberg AFB and the California

Department of Forestry (CDF) have been sharing GIS data and analyses about brush fires in both jurisdictions. The CDF and Vandenberg AFB firefighters used the data and information to improve their joint firefighting.

Such I&E geospatial data asset uses are especially important for improving communications in processes where many different organizations are involved, such as in SERPPAS (see the discussion in Chapter Three). The maps created by DISDI for the SERPPAS collaboration improve the communication processes among OSD, southeastern installations, Florida, Georgia, South Carolina, North Carolina, the Conservation Fund, the Nature Conservancy, and other stakeholders in this collaborative sustainability process. Seeing different organizations' areas of interests on the map, such as installation encroachment and urban sprawl concerns, helps everyone to understand and recognize areas of common ground and mutual benefit.

Other Process Changes. The use and sharing of I&E geospatial data assets also cause other types of process changes, including analysis, organizational, and decisionmaking process changes. In many cases, the use of I&E geospatial data assets computerized part or all of a process, i.e., automating a process within the computer that was done by hand before, such as in the Keesler dig permitting process, which no longer required that someone drive the dig permit all around the base.

Often, there was also a change in the analysis or decisionmaking process itself. Such changed processes often involved using more precise geospatial data, being able to integrate more diverse data from different sources, and assessing the data within the integrated computerized system, with the information at the users' fingertips in a spatial form. For example, at the installation level, I&E geospatial data assets have changed the facilities planning analysis process. At NAS Patuxent River, the installation GIS was used to assess staff dining facility needs and to determine the best location for a new cafeteria. In 1998, when several thousand new personnel were being added to the base because of the 1995 BRAC, many of the new employees were commuters from the District of Columbia and they needed cafeterias within walking distance of their employment locations. The installation was adding 2.0 to 2.5 million square feet of office space for the new staff and needed to place a new cafeteria. The GIS shop analyzed the existing eating loca-tions and created buffers, showing the number of people who would now be nearby and where they might need a new cafeteria.

There can also be a broader organizational process change. For example, Camp Butler Environmental Management staff members are building an integrated and comprehensive GIS model of Okinawa to use for the Camp Butler Environmental Management System (EMS). There is the potential to have a significant process change regarding environmental management. Since this system involves linking all the various environmental databases to the GIS, there is the potential to change how environmental issues are managed at the base, evolving to a more accurate place-based system. If an incident, planning issue, or concern occurs at any location on the base, one could immediately bring it up in a GIS and look at all the environmental concerns— whether storm water runoff, endangered species concerns, or pesticide application.

Other Mission Effects

Finally, there are a host of other mission effects from using and sharing geospatial data assets. These effects are often specific to the organiza-tion and the way the geospatial data assets are used. The dig permitting application at Keelser AFB, besides having efficiency and effectiveness benefits, also has employee morale benefits. Not having to hand-carry a dig permit all around the base for approval helps reduce employee frustration. For discussion purposes, we group the rest of these other mission effects into four main categories:

- changes in policy
- educational and training effects
- public relations effects
- legal effects.

Changes in Policy. Policy effects occur when an organization changes its policies as a result of having and using geospatial data assets. The effects may result from changes in analytical processes. Perhaps the best example of a policy change is the use of IVT data as official data during the BRAC process. According to an April 2003 memorandum from the Chairman of the Infrastructure Steering Group that headed the 2005 BRAC process in the Office of the Under Secretary of Defense for Acquisition, Technology, and Logistics (OUSD(AT&L)), IVT was a "planned capability to enhance the Department's overall ability to manage its infrastructure . . .[that would] assist the JCSGs, the IEC [Infrastructure Executive Council] and ISG [Infrastructure Steering Group], and DoD Components in their BRAC 2005 analyses."[9] This was the first BRAC process to use I&E geospatial data as certified data subject to audit by the Government Accountability Office (GAO).[10]

Installation policies have also changed from the use of I&E geospatial data assets. The April 2005 range fire in the CTA at Camp Butler, discussed above, created a large amount of political attention for the USMC because of the erosion concerns on Okinawa. Initially, USMC Camp Butler Public Affairs stated that the fire covered an area 1 km by 1/2 km, according to the longest length and width, not accounting for the fact that the burned area was not a true rectangle. The burned area was actually one-third this size. Using GIS, the true burn area was calculated and the revised data were released to the public. Giving two different numbers about the acreage burned initially hurt USMC credibility and now Camp Butler has a new policy that Public Affairs will not release burn size numbers until they have been calculated in the GIS. This incident shows how the GIS implementation helped change an official base policy, causing it to use more accurate information from a GIS calculation in public relations announcements. Besides explaining the fire extent, the geospatial analyses were important in helping explain the erosion and erosion control efforts to the local public.

Educational and Training Effects. Education and training effects are seen where the use of I&E geospatial data assets requires more or less education and training to use them. This effect can be both posi-

---

[9] Aldridge (2003).

[10] Effective July 7, 2004, this agency's name changed from General Accounting Office to Government Accountability Office.

tive and negative. For example, using a new geospatial application may require some additional training. For example, the IVT program office provided three- to four-hour briefings and training sessions to the JCSGs on how to use the IVT Viewer.

In other cases, an organization may be able to save education and training time by providing easy-to-use geospatial applications, such as web viewer systems. For example, the NAS Patuxent River RSIP Viewer provides mapping support for running of the base where users do not need any knowledge of GIS. Specifically, the portal allows many diverse users to access I&E geospatial data assets for diverse business functions. This portal has about 900 active users who log on to the system. Many of these users use standard map products, such as customized base maps for different applications, including air operations, public safety, and environmental and facility support. In 2002, about 1,300 map products were provided through RSIP; in 2005, 1,009 map products were supplied (for more details, see the appendix).

In such cases, more GIS and web applications development training may be needed for the staff who developed this system. However, even these training costs and education can be offset. For example, the environmental manager at Camp Butler has hired environmental staff with GIS skills to leverage resources, so that he can experience the development of sophisticated geospatial applications without investing in GIS training.

Public Relations Effects. I&E geospatial data assets can help improve relations with the general public, other federal agencies, foreign governments, and state and local governments near installations through improved management of processes and improved sharing of information. The latter is closely related to the improved working relationships discussed above. One way in which I&E geospatial assets affect relations with the general pubic is through management and outreach about environmental and cultural resource issues. For example, NAS Patuxent River received complaints from community members regarding UAV operations over the Northern Neck of Virginia (see the discussion in Chapter Three). Installation GIS staff used GIS to determine alternative UAV routes in an effort to lessen the effects on the Northern Neck of Virginia community. Staff members showed the maps of this analysis and new routes at public meetings. The use of the maps helped eliminate noise complaints from residents. Similarly, I&E geospatial data assets are used routinely to help explain to the public about NEPA processes and in the development of INRMPs.

Sharing of I&E geospatial data assets because of safety, emergency response, and environmental concerns, such as encroachment, has helped improve relations with local governments, and environmental groups in the case of environmental data. For example, relationships with state and local governments and environmental NGOs were improved in the ACUB process because of I&E geospatial data sharing (see the ACUB discussion in Chapter Three).

I&E geospatial data assets have also been used to help improve public relations with Congress. Both USAF and Army headquarters organizations have used these assets to support analyses and provide maps to Congress, as discussed more in the appendix.

Legal Effects. The use and sharing of geospatial data assets can have several types of legal effects. For example, I&E geospatial data can be used to fulfill legal requirements, such as in the NEPA process; they can be used as evidence of wrongdoing; and they can be used to determine legal liability for accidents and other harm that may occur. We provide examples of the latter two cases.

I&E geospatial data assets support installation security by providing legal evidence about traffic incidents or crimes on base or about installation boundaries, such as violations by fishermen and hunters who are caught on a military installation and claim they were not on the base. For example, at Marine Corps Base Camp Pendleton, California, geospatial staff members have used I&E geospatial data assets to provide traffic accident maps as evidence for the military police to use in court.

After the April 2005 fire at Camp Butler, Okinawa, local fishing communities complained that soil erosion from the fire had hurt their fishing grounds. The local community commenced a claims process against the U.S. military seeking compensation. Camp Butler environmental staff used the 3-D GIS data to show that the fire did not go over a ridge, so there was no erosion on the other side of the island. Thus, the fishermen had no basis for claiming compensation.

Multiple Effects

We have given examples of different types of effects, from those easy to measure, such as time savings, to those much more difficult to measure, such as improved public relations. But, in fact, each time I&E geospatial data assets are used or shared, multiple benefits accrue, as mentioned in the examples above and in Chapter Five's discussion of the use and sharing of IVT data. As discussed there, not only did the IVT data and process help improve decisionmaking and communications and save time in the BRAC process, they were also instrumental in the development of Service GIO I&E geospatial data programs and their QA/QC processes. When examining an application, it is important to examine the full range of effects, which also can include negative ones, such as the cost for updating and maintaining the geospatial application itself, and effects on other organization and mission areas; this also was demonstrated well by the IVT discussion in Chapter Five.

Here, we present two examples to show the wide range of effects that the use of I&E geospatial data assets can have and how important it is to look at the full range of effects. The first example involves parking space management at an installation and the second relates to OSD critical infrastructure protection. Table 7.2, at the end of this section, highlights some examples of multiple effects for diverse applications.

At USAF's Aviano Air Base in Italy, I&E geospatial data have been used to help assess and allocate parking spaces. This may seem like a very specific and narrow application related to an installation management function. However, the development of the data for this application has had many other effects in other areas, including other mission functions. Because of USAF requirements about parking, the base commander at Aviano Air Base asked the Civil Engineering staff to assign parking on base to meet the requirements regarding open and reserved parking. Because the details about parking spaces are not in the USAF CIP, the Aviano Air Base GeoBase staff used GPS to record locations and to help digitize the parking spaces and then entered attribute information about the spaces into the database. Attribute information included the types of spaces, such as private or government-operated vehicle reserved, or open space and American Disability Act spaces. The system is used to more effectively allocate and manage parking spaces and has helped reduce the base parking footprint. The system can also be used to help calculate contingency parking for special events and incidents.

Having parking space data has many other advantages besides parking space management. First, the data help with parking enforcement and parking lot maintenance. Security forces are now given maps with detailed information about parking types to use for ticketing purposes. Maintenance staff members now know precise area measurements for paving and lineal feet measurements for paint striping so they can more effectively manage contractors who perform these functions, such as telling the contractors how much paint they will need to paint the parking lots. Second, parking lot geospatial data are also useful for both AT/FP analyses, such as assessing standoff distances from parking areas, and for explosive safety distance assessments, such as the information being used in ASHS. Third, parking data also have been a communication tool to base staff, by giving them an accurate picture of legal parking. Last, parking data also help those responding to emergencies.

There are some less-beneficial effects of this parking space system, namely, the manpower cost to create and maintain it. Entering the data is manpower-intensive and, as parking is reallocated, manpower is needed to update parking space information changes in the GIS. Such manpower costs can be reduced in several ways, such as by using recent aerial imagery to help create or update the parking space data. Also, procedures can be developed to streamline updates. This simple example shows how the development of seemingly mundane geospatial data may have an upfront investment cost in manpower but has many mission benefits once they are used and shared for multiple purposes.

The second example relates to how the Office of the Assistant Secretary of Defense for Homeland Defense Critical Infrastructure Protection wants to use I&E geospatial data assets. This OSD office conducts Critical Infrastructure Protection mission area analyses at installations to assess vulnerabilities in assets that support the mission. This process involves teams of 10–15 people traveling to each base. It costs $300,000 per base for each assessment. A potential process change would be to use I&E geospatial data where a preassessment is performed using geospatial data and then to focus on the gaps.

Teams then would need to visit only bases where gaps exist and fewer people would be required. Antwane Johnson, OSD, estimates that this approach would result in 20 percent to 50 percent savings per base, for an estimated total cost savings of $6 million to $15 million per year for 100 base assessments. He presumes that sufficient GIS data are available for all 100 installations so that travel would not be necessary. If this were not the case, savings would be less. However, regardless of the exact assumptions, time and manpower would be saved as a result of the new analytical process.

Another analytical process change could affect how OUSD(AT&L) makes resource decisions. If the critical infrastructure vulnerability assessment results were integrated into an I&E geospatial system so that the information reached OUSD(AT&L) decisionmakers, they would see vulnerabilities that they could fund in their resource allocation process. Thus, this application could change an analytical process, help with OSD critical infrastructure protection analysis and installation resource allocation decisionmaking, and save money, manpower, and travel requirements.

Table 7.2 demonstrates the varied and widespread effects of using and sharing geospatial data, and how a single application can have multiple effects. The information in Table 7.2 draws from discussions above, in previous chapters, and in the appendix. This table illustrates the main type of effects for different sample uses of I&E geospatial data assets. As was just demonstrated with the parking space management and critical vulnerability assessment examples, the actual and potential effects of using and sharing I&E geospatial data assets are complex, and the illustrations in the table cannot capture these complexities. Each example would need to be more fully examined to understand and assess the full range of possible effects; the next section presents a methodology for doing just that.

However, in general, this table makes clear that there is often an upfront cost to develop and apply the I&E geospatial data assets but that there can be significant time savings overall. In addition, the upfront costs are reduced when geospatial data and software applications are shared. The table also shows that there usually are communication benefits and, for some uses, benefits to multiple missions.

Table 7.2
Sample Mission Effects for Sample I&E Geospatial Data Asset Uses

Mission Area	Sample I&E Geospatial Data Asset Use	Sample Efficiency Effects	Sample Effectiveness Effects	Sample Process and Other Effects
Base management and operations	Keesler AFB dig permitting GIS-based process being used to process dig permits	Initial staff time spent to develop the tool. For each permit, saved over one-half the time in dig permit processing. Helped avoid costly accidents of hitting utility lines	Dig permits are more accurate and processed faster	Improved understanding and communication among different staff; improved worker morale
	Aberdeen Proving Ground using GIS to develop options and assess where to place automotive testing track	Saved over 10 weeks staff time to develop and process four options	More accurate view of all the building constraints; more effective decision about siting the testing facility	Improved communications and relations at working group meetings to review the siting options; changed the analysis process
	Aviano Air Base using GIS for the allocation and management of base parking spaces	Manpower intensive to first develop, plus some initial time needed to update and maintain	Reduced base parking footprint; improved contractor management; parking space management and enforcement; AT/FP analysis; explosive safety assessment; and emergency response	Improved communication and outreach about parking space locations and policies
Emergency planning, recovery, and response	Using GIS to develop spill response plans at Naval Support Activity (NSA) Capodochina, Italy	Environmental staff time to develop and apply the GIS tool	Improved spill emergency response, planning, and operations, including faster response times; improved spill control and postspill analysis and cleanup	Improved communications at command and control meetings; short amount of time needed to train users about the tool

Table 7.2—continued

Mission Area	Sample I&E Geospatial Data Asset Use	Sample Efficiency Effects	Sample Effectiveness Effects	Sample Process and Other Effects
	Langley AFB Floodmap tool being used for hurricane preparedness and response	Initial contractor cost to develop the tool; many fewer hours of GIO staff time to develop impact scenarios and produce maps	Improved preparedness, planning, response, and situational awareness; faster response times and more accurately placing resources where needed during the emergency	Improved communication and working relationships among different functional staff responding to hurricanes
Environmental management	Camp Butler 3-D watershed modeling and burn analysis to reseed burn area	Extra staff time to do the GIS analysis and perform hydroseeding operations	Reduced erosion and helped restore vegetation	Changed policy for release of burn information; helped relations with Japanese government; helped show reduced legal liabilities
	Patuxent River NAS using GIS to develop new UAV routes because of noise complaints and presented at public meetings	Initial staff time for the GIS analysis, but less staff time spent on noise complaints	Improved operational safety and fewer noise complaints at the base	Helped public relations and to educate UAV operators about community concerns
	DISDI providing installation GIS data and an OSD contractor performing GIS analysis and maps to help OSD and southeastern states work together in SERPPAS	Manpower cost and time to do the analysis and develop the maps	Increased meeting productivity; provided integrated and common view of the areas of concern; helped the collaboration identify potential areas of focus within the region	Helped improve DoD collaboration with state and local governments and environmental NGOs; helped different organizations understand each other's needs and common interests and goals

Table 7.2—continued

Mission Area	Sample I&E Geospatial Data Asset Use	Sample Efficiency Effects	Sample Effectiveness Effects	Sample Process and Other Effects
Homeland defense, security, and critical infrastructure protection	OSD critical infrastructure vulnerability assessments	Fewer bases to be visited, saving travel costs and staff time	Potential for improved critical infrastructure protection and infrastructure resource allocation	Potential to change the analysis process; potential to integrate results into OSD resource allocation systems to make the information more accessible
Military health	USAF Surgeon General Modernization Directorate using CHaMP to track influenza-like outbreaks	Manpower cost to develop, maintain, and run CHaMP	Improve tracking and assessment of avian flu and other influenza-like outbreaks; potential to speed the efforts of stopping the spread of the disease	Potential to help improve communications about influenza outbreak
MWR: enhancing quality of life	Using GIS to help plan July 4th concert on the beach at Camp Pendleton	Staff time to do GIS analysis and produce maps; includes developing emergency response scenarios for military police and other first responders	Improved event operations, including traffic and parking management, and security and emergency response procedures	Improved communications among MWR, police, and emergency response staff; helped improve PR information about the event
Safety and security	Use of ASHS to assess capacities for explosive safety at Osan Air Base in the Republic of Korea	Contractor cost to apply ASHS model and do the assessment	Ability to store more munitions; improved flow of munitions in storage area and flight line; improved flow of sorties	Improved base GIS data quality by validating, adding to, and updating them

Table 7.2—continued

Mission Area	Sample I&E Geospatial Data Asset Use	Sample Efficiency Effects	Sample Effectiveness Effects	Sample Process and Other Effects
	GIS analysis to reduce BASH at Langley AFB	Manpower for GIS analysis	Decrease in air strike hazards because fewer ospreys near runways; osprey population remained stable	Public relations benefit because of protecting the osprey
Strategic basing	Use of IVT data to help support the BRAC process	Service and IVT Program office manpower time and cost to develop the IVT data and tool; however, saved time for BRAC analysts in their processes and the production of PowerPoint charts and maps	Improved decisionmaking and provided common situational awareness; numerous maps generated for OSD senior decisionmakers in short amount of time—a service that could not have been done before	Improved communications with senior leaders and Congress; JCSG staff time for minimal training on how to use the tool
Training	Applying RMTK for range design and planning at Army bases	Initial investment cost to develop the tool and apply it; faster process for assessing range effects	Improved assessment of range effects, including noise and safety; improved siting of the range; reduced environmental and safety effects from the range	Improved communications and working relationship among range planners and operators and the environmental staff

Table 7.2—continued

Mission Area	Sample I&E Geospatial Data Asset Use	Sample Efficiency Effects	Sample Effectiveness Effects	Sample Process and Other Effects
	GIS analysis to help Camp Lejeune reconfigure a range away from a river	Minimal GIS staff time needed to perform the analysis	Improved range reconfiguration; improved assessment of range effects, including safety and environmental; helped minimize effect on T&ES and wetlands; helped with the NEPA process	Improved communications between environmental and range staff; supplied maps for legal NEPA documents and public outreach process
	Use of USAREUR ITAM Mapper for training planning and operations	GIS staff time to develop and maintain tool, but time savings in not having to produce so many maps; savings in map printing costs	By proving more timely and accurate maps, improved training planning and operations, including soldier training orientation; improved environmental awareness	User-friendly interface saved educational time; improved communication for joint training

Our Methodology for Evaluating Effects

In the previous sections, we listed and classified the different effects on organizations that use and share geospatial data assets. We now turn to the question of how to assess, evaluate, or measure those effects. Most of the measurements cannot be highly quantitative because, as we have shown in this chapter, many effects are difficult to quantify. However, here we present our methodology, which uses three complementary approaches; in combination, they capture the full range of effects and exploit available data:

1. an information flow model to understand the range of organizations using and sharing an I&E geospatial data asset
2. a set of logic models to map out how the inputs, activities, and outputs of an organization's data development, use, and sharing lead to outcomes for different customers
3. to the extent possible, when the data are available, employment of a variety of methods for quantifying the logic models.

We discuss each of these parts of the methodology below and present some illustrative examples.

Information Flow Model

Our methodology calls for an examination of the range of organizations that share or use a single set of geospatial data or an application. We call this an information flow model because it demonstrates how geospatial information passes from one organization to another. This is the first step to understanding how geospatial data assets are shared. Along the way, each organization may see one or more of the effects we presented above.

Figure 7.2 traces the flow of geospatial information used for IVT data creation in the BRAC process. It also shows the official BRAC users of IVT data. In this process, the data flow from U.S. military installations all the way up to the Secretary of Defense and Congress. The boxes represent organizations, the circles represent data, the ovals

Figure 7.2
Information Flow Model for IVT Data in the BRAC Process

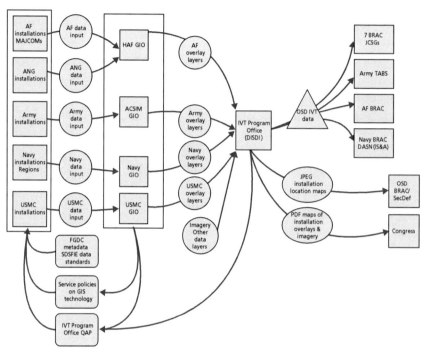

represent applications or uses of the data,[11] the triangle represents a data repository (the IVT itself), and the rounded boxes represent standards and policies. Figure 7.2 shows that the flow of geospatial data assets in IVT began with standards, policies, and quality assurance plans specified by the IVT program office applied by the separate military installations to geospatial data collected and managed by the installations (in fact, the data in most cases were collected and managed by numerous separate offices and directorates at each installation). Once the installation commanders validated the data, they were submitted to their respective Service offices. Each Service office compiled the geospatial

---

[11] These applications, combined with the data, are what we refer to as geospatial data assets.

data and performed addition quality assurance before submitting the data to the IVT program office (now DISDI). The IVT program office then combined that Service data with other geospatial data, including imagery. The combined geospatial data assets became the official OSD IVT data that were shared with seven Joint Cross-Service Groups as well as the Army, Navy, and Air Force BRAC offices as part of the BRAC process. Because of OSD policy and licensing restrictions on how the IVT data could be shared, only installation maps with imagery overlays in PDF format were shared with Congress; the Office of the Secretary of Defense, however, requested numerous installation location maps, which were provided in JPEG format.

Figure 7.3 shows how some of these data were shared for uses outside the official BRAC process. It completes the geospatial information loop, with data and imagery passed back to the Services as appropriate. Once DISDI had the official DoD installation data, it shared Service-specific data back with the Services. In some cases, the Services combined those data with additional imagery, Service data, or new data for use in an application. For example, ANG civil engineers used ANG IVT data, combined with USGS imagery and other data, to develop parking plans for ANG aircraft under a variety of BRAC scenarios. The Army Corps of Engineers used Army IVT data, again combined with new and additional data, to help conduct land-use change analyses at 100 Army bases to examine encroachment issues from surrounding communities.

Ultimately, these applications were shared with installations, individual generals and their staffs traveling to the installations, trainers, BRAC offices, and legislative liaisons. Along the way, the organizations and individuals who used and shared the geospatial data assets from IVT would have experienced many of the effects presented above, from time and cost savings to improved working relationships and policy effects.

Logic Models

The second step in our analysis process is to use logic models. The logic model is a versatile analytical tool that has been in widespread use for years. Government agencies, private foundations, and businesses use

Figure 7.3
Sharing of IVT Geospatial Data

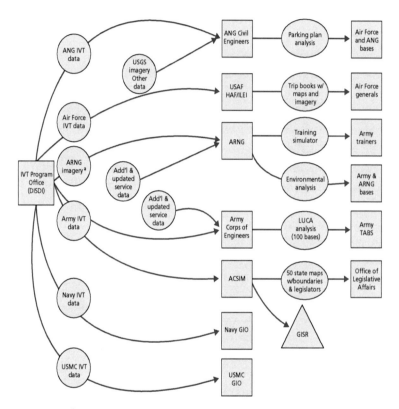

NOTES: This figure shows only some of the ways IVT data were used outside the BRAC process. For example, Navy and USMC GIO examples are not included in this figure. Chapter Five discusses some of these additional examples. For a full assessment of the IVT program's effect, the additional examples would need to be modeled and examined.
[a]Imagery for ARNG sites was acquired for the IVT process but was not included in final data of IVT product.
RAND MG552-7.3

logic models to conduct mission effect assessments across a wide range of programs and policy areas, including public health, education, agriculture, research, social work, and technology transfer. Logic models illustrate how the inputs and activities of an organization potentially lead to beneficial outcomes—in other words, logic models illustrate the underlying logic of an organization's activities.

Figure 7.4 presents a "textbook example" of a logic model, as presented in the W. K. Kellogg Foundation's *Logic Model Development Guide*.[12] The basic model is flexible enough to include greater detail or different factors shaping the path from inputs to outcomes. For example, logic models might divide outcomes into short-, intermediate-, and long-term outcomes or they might explicitly state assumptions, external conditions, or strategies. Logic models might also include various performance metrics.

Our logic models are slightly different from the notional model of the W. K. Kellogg Foundation, but they are founded on the same principle of mapping inputs and activities to outputs. Our logic models differ in three key ways, based on examining previous RAND research and federal agency experience, such as EPA and DOE, for applying

Figure 7.4
Textbook Example of a Logic Model

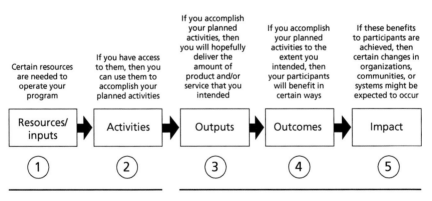

SOURCE: W.K. Kellogg Foundation (2004).
RAND MG552-7.4

---

[12] We offer this as a textbook example because a number of organizations reference it. See W. K. Kellogg Foundation (2004).

logic models to government processes.[13] We briefly discuss the three differences here and then demonstrate them as we apply the logic models. First, our logic model includes customers because there are different DoD and other organizations' "customers" that benefit from the development and use of I&E geospatial data assets. Second, we do not include effects as separate boxes in the models, since an effect is a contribution to an outcome. In addition, I&E data use and sharing can also have an effect along the operational path in a logic model. Third, our models discuss actual rather than proposed outcomes.

Before presenting examples of logic models, it helps to set the context by examining the potential for the proposed methodology for assessing the effect of I&E geospatial data asset use and sharing. The information flow model and the logic model provide different perspectives on effects. The information flow model shows how information is shared and with whom, very similar to DoD's Business Enterprise Architecture information exchange analysis. The information flow model may, therefore, support larger efforts within the Business Management Modernization Program within DoD. The information flow model illustrates where DISDI and other organizations fall within the process of sharing geospatial data assets; by contrast, the logic model focuses on a single organization, such as DISDI, and its inputs, activities, outputs, customers, and overall outcomes. Logic models can show policy effects (via changes to inputs and activities), effectiveness effects (such as new or better outputs), and even quantifiable efficiency effects. Customers in one logic model can have a logic model of their own. By "daisy-chaining" the logic models together, one can get an even better picture of the effects of using and sharing geospatial data assets. DISDI and other organizations can use this approach to help understand, assess, and explain the full range of effects from their use and sharing of I&E geospatial data assets.

Now we will present logic models that have a range of activities, outputs, customers, and outcomes, and then we will present logic models that actually have metrics to measure the effect of using and sharing geospatial data assets. The different logic models we present

---

[13] For such logic model examples, see Greenfield, Williams, and Eiseman (2005).

illustrate different points relevant to the effects of sharing and using geospatial data assets. First, we present two logic models of the IVT process. Then, we present some installation logic model examples.

IVT Program Office. We begin with a rather simple model that traces the inputs, activities, outputs, customers, and outcomes of the IVT program office. Second, we trace the effects further downstream through one of the JCSGs that was a customer in the first part of the model. To fully access the effect of the IVT data's use and sharing, many additional logic models would be needed for all the different organizations involved. The two examples here seemed sufficient to show how to apply this methodology to this complex and widespread data-sharing process. In addition, the flow model presented in the previous section would provide guidance in developing additional logic models to represent other linkages between the IVT program office and customers downstream, and the IVT details discussed in Chapter Five could be used to help provide the detail in these models.

The inputs to the IVT products included a variety of geospatial data provided by the installations (IVT submittal packages), imagery and other data from noninstallation sources, plus policies and standards (from OUSD(AT&L), the IVT working group and the IVT technical group). Inputs also include labor, i.e., the DISDI staff time to perform the activities. The IVT program office then conducted a number of activities, including data assembly, quality assurance/quality control, training, software modification, and writing, that resulted in several different types of outputs. As discussed in Chapter Five, the outputs went to different customers: PDF installation maps to Congress, JPEG installation location maps to the Secretary of Defense and the OSD BRAC office, the OSD IVT system itself (the IVT geospatial data plus the IVT Viewer on a laptop) to each of the seven Joint Cross-Service Groups (Education and Training, Headquarters and Support Activities, Industrial, Intelligence, Medical, Supply and Storage, and Technical), and the Service IVT data to each of the Service BRAC offices. In addition, the activities of the IVT program office also generated data outputs that the Services were able to use in a variety of applications, demonstrated in the flow model in Figure 7.3.

Figure 7.5 shows the logic model for the IVT program office production of IVT products. This logic model shows how the use of the IVT system and related data led to effectiveness effects, specifically, improved decisionmaking in the BRAC process, by providing situational awareness; helping support other analyses, such as combining IVT data with other data in the military value calculation; helping show relationships, such as encroachment concerns; as well as communication benefits, such as helping to explain decisions to senior leaders. It would be extremely difficult to try to attach a dollar value or to otherwise quantify the contribution of the IVT system and related data

Figure 7.5
Logic Model for the IVT Program Office's Production of IVT Products

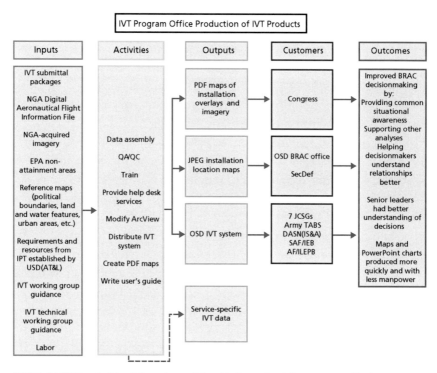

NOTE: SAF/IEB = Assistant Secretary of the Air Force for Infrastructure, Environment, and Logistics/Basing and Infrastructure Analysis; AF/ILEPB = Civil Engineer of the Air Force, Programs Division, Bases and Unit Branch.
RAND MG552-7.5

to the quality of the decisions made. However, our interviews with members of the different Joint Cross-Service Groups, plus the number of maps that were created for the OSD BRAC office and the Secretary of Defense, suggest that decisionmakers valued the geospatial data assets quite highly. This logic model also shows some efficiency effects in the form of more quickly producing maps and PowerPoint charts without as much manpower. Given the complexity of this process, this logic model captures only some of the effects. To fully capture them all, a logic model would need to be developed for each customer to understand how each one used the IVT information and its effect on each customer's outcomes. Next we present such a logical model for the Medical JCSG.

What does improved decisionmaking actually mean in terms of effects for each customer in this process? The logic model for the Medical JCSG illustrates what improved decisionmaking meant for this organization (see Figure 7.6). Recall that the IVT Viewer and data were provided to each of the seven JCSGs during the BRAC decisionmaking process. These were the customers of the IVT program office, and their use of the data supported BRAC decisionmaking. The Medical JCSG used the IVT data to support analyses that examined and recommended sites for new construction and for conversion of existing facilities. This ultimately led to better location of medical facilities, time savings, dollar savings, and greater confidence in the decisions that were made. Again, for more detailed discussion of this process, see Chapter Five.

Camp Butler Environmental Management Program

At the installation level, one organization can share data for many purposes that support multiple mission areas and that sharing can have a broad range of effects, as illustrated by the next two logic models for Camp Butler's Environmental Management Program (see Figure 7.7). The logic model for Camp Butler's Environmental Management Program shows that the program uses data that are collected and managed on the installation as well as data provided by the local community. One of its key intermediate activities is the creation and maintenance of a 3-D model of Okinawa. This model supports watershed

Figure 7.6
Logic Model for the Medical JCSG

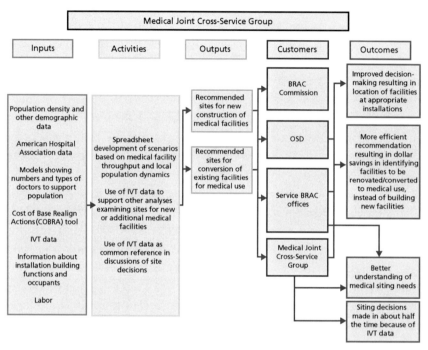

RAND *MG552-7.6*

modeling, training range development analysis, tidal wave simulations, and creation of a 3-D video by the Environmental Management Program. Environmental staff members use their 3-D model and GIS analyses to create a range of outputs for different mission functions and ultimately for different customers, which leads to different types of outcomes from sharing and using geospatial data. Some of those effects were expected and others, most likely, were not. The Environmental Management Program at Camp Butler probably did not invest in geospatial technologies with the expectation that they would change the policies of the public affairs office, but that is what happened after an accidental fire occurred on Okinawa from training operations at the installation. This changed policy has resulted in more accurate burn information being released publicly (see Figure 7.7). Figure 7.7 also

Figure 7.7
Logic Model for Camp Butler Environmental Management Program's
Production of I&E Geospatial Data Products

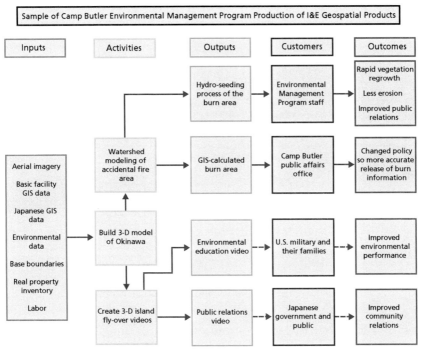

RAND *MG552-7.7*

shows how the 3-D video is supposed to help educate USMC troops and their families about environmental concerns, such as the storm water runoff sensitivities and the natural and cultural resources of Okinawa. With this knowledge, they can modify their behavior so as to have less effect on the environment, resulting in improved environmental performance and protection. The dashed line in the figure indicates that these are planned outcomes, since, at the time of our study, the video was a prototype that had not yet been used for these planned purposes.

The logic model for part of Camp Butler's environmental management program in Figure 7.7 shows that the program uses geospatial data that are collected and managed on the installation as well as

Figure 7.7—continued

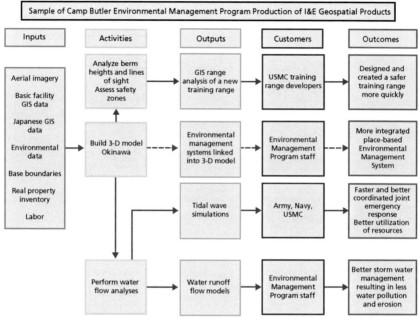

RAND MG552-7.7, continued

data provided by the local community. Again, one key intermediate activity is the creation and maintenance of a 3-D model of Okinawa. The activities and outputs of this program support different customers and mission areas (listed under outcomes), including training, emergency response, and environmental management. For example, the GIS training range analysis of assessing different berm heights and their effect on training range lines of sight and of safety zones resulted in more quickly designing a new and safer training range; the tidal wave simulations improved joint planning and emergency response training, which results in better coordination and communication, faster response times, and better use of resources for an emergency incident; and the water runoff flow models and analysis helped the environmental staff better manage storm water runoff so that there are fewer pollution and erosion problems, such as more efficiently placing the technologies to capture and treat the oil runoff from parking lots.

The dashed line indicates that one output based on the 3-D model is planned but has not been implemented.

To track effects over time, the Camp Butler environmental management program or some other group studying the program's effects could track the number of outputs that are created using the 3-D model, the number of customers that use one of the outputs, and the range of outcomes from their use. The outputs in these model fall into all four of our categories: effectiveness effects (vegetation regrowth and less erosion; improved environmental performance; improved emergency response), improved communications and working relationships, public relations effects, and policy effects. If it can be determined that the program saves time or resources in producing the current outputs or new outputs, we will show in later logic models how those resource savings can be demonstrated and even measured.

It is important to note that this logic model shows only a small sample of how the Camp Butler environmental management program uses I&E geospatial data assets. Additional logic models would need to be developed to show the full range and effect of their activities.

NAVAIR Range and Sustainability Office

The next logic model highlights the many customers who use different outputs that are created using geospatial data assets (see Figure 7.8). By providing the outputs to those customers, a variety of outcomes is realized. The outcomes fall into several categories from our taxonomy: improved decisionmaking, working relationships (with the contractors), performing a new task that could not be done before, and public relations effects. The logic model shows all the inputs and activities that led to those outputs.

The activities are conducted by the Range and Sustainability Office, but the inputs themselves came from a variety of sources. In other words, they are geospatial data assets that were shared by some groups and ultimately used by others. As was the case with several of the other programs discussed in this section, some of the geospatial data used by the Range and Sustainability Office came from nonmilitary sources, in this case from the local community, which in turn was also a customer of the office's outputs. The sharing of geospatial

Figure 7.8
Logic Model for the NAVAIR Range and Sustainability Office at Patuxent River NAS

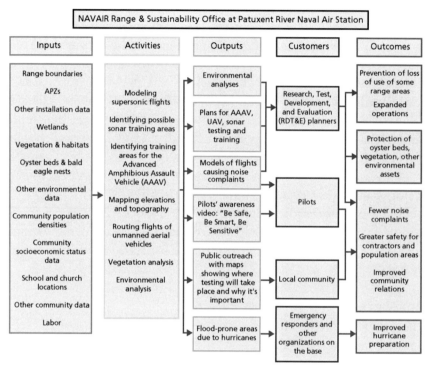

RAND MG552-7.8

data assets again resulted in many effects, as discussed above. Those include education and training effects, improved operations, public relations effects, improved working relationships, and the ability to perform a new task that could not be done before.

Quantitative Methods for Evaluating Effects

The third part of our methodology is to quantify effects when data are available. First, we show how this can be done in conjunction with the logic models, by showing manpower savings for the Langley AFB Tank Management Program and the construction office. In other cases, more complete data are available because other researchers have conducted a benefit-cost analysis or related study. Second, using these quantitative measures, we show how a relatively straightforward extrapolation can be done to get a rough order-of-magnitude estimate of the total potential effect across all DoD installations.

Langley AFB Tank Management Program. We now show how some of the effects in logic models can actually be quantified and even valued. In Figure 7.9, we show the logic model for Langley AFB Tank Management Program production of dig permits before the use of installation I&E geospatial data assets in the top panel and the model after the use of Langley GeoBase system, i.e. the Langley Geo-Base MapViewer, in the bottom panel. The Langley AFB Tank Management Program produces dig permits that are required any time a

Figure 7.9
Logic Model for Langley AFB Tank Management Program Production of Dig Permits

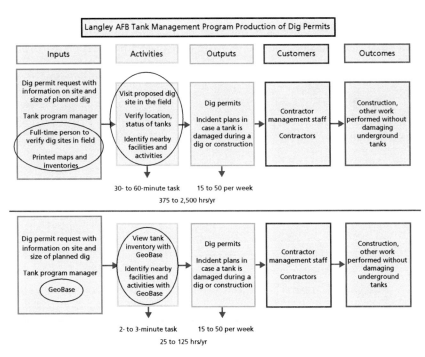

construction project will be turning earth. Without the Langley Geo-Base MapViewer (see the discussion in Chapter Two), the Tank Management Program would require a full-time person to visit proposed dig sites in the field using printed maps and inventories that had to be updated every time an infrastructure change was made. The person would have to travel to the site in the field, verify the location and status of each tank, and identify nearby activities and facilities. This could take 30 to 60 minutes for each dig request submitted to the program. The program would then make a decision about approval of the dig request and would develop incident plans in case of damage to a tank during construction.

With 15 to 50 requests to process each week, the total amount of staff time spent on site visits could range from 375 to 2,500 hours per year. Using the installation web-based GIS system, someone in the program simply views the tank inventory and identifies nearby activities and facilities electronically, and the process takes only two or three minutes. The total amount of time required to perform these activities could range from 25 to 125 hours per year. It is important to note that other installations also had similar dig permitting tools, such as Keesler AFB, which had similar or even larger man-hour time savings from the use of its I&E geospatial data application.

Langley AFB Construction Office. The construction office at Langley AFB also incorporates geospatial data into its business processes for the production of delivery orders. The logic model for this office is shown in Figure 7.10.

Using logic models, we have illustrated a number of effects from using and sharing geospatial data assets, ranging from the very broad (improved decisionmaking) to the very specific (time savings). We showed how manpower savings could be quantified within a logic model. We now turn to another method to help quantify effects, benefit-cost analysis.

Using the Langley GeoBase system, the construction office generates drawings showing the area of the work, utilities in the area, and haul routes for the construction materials. This saves between 100 and 225 hours per year generating delivery orders for construction contracts. The two construction-related functions—dig permitting and

Figure 7.10
Logic Model for Langley AFB Construction Office's Production of Delivery Orders

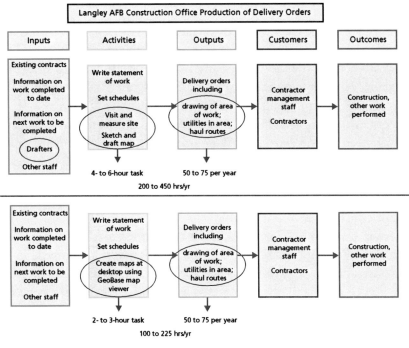

RAND MG552-7.10

construction order generation—save 450 to 2,600 man-hours per year at Langley AFB. Similar manpower savings calculations could be made for other installation functions from other bases applications, such as GIS-based tools for looking at environmental and other constraints to help select a location for building a structure on a base.

Benefit-Cost Analysis

Benefit-cost analysis is very simple conceptually and something everyone does every day, albeit in a less rigorous manner than an economist would when studying IT investment decisions. A benefit-cost analysis adds up all the gains from each alternative, subtracts the costs, and selects the alternative that maximizes net benefits. Variations on benefit-cost analysis include return on investment, which calculates a ratio of benefits to costs, and cost-effectiveness analysis, which compares the relative costs of generating a desirable outcome.

These methods are conceptually simple but can become computationally complex, with benefits and costs accruing at different times and with different levels of certainty or risk. They are also data-intensive and not easily updated. But perhaps the biggest limitation of these methods is the fact that many benefits are difficult to measure, let alone value. Improved decisionmaking is one such benefit. Estimating the economic value of improved decisionmaking is a two-stage process: estimating the benefit of the decision and then estimating the contribution of the geospatial data assets to the decision. This yields values that are typically very large relative to other quantified benefits and costs, and so experts recommend reporting such estimates separately rather than including them directly in the computation of ROI or cost-effectiveness.[14]

The USGS has developed a "GIS benefits estimation model" that predicts the economic value of benefits that may initially appear to be time-consuming, expensive, or even apparently impossible to quantify. The model separately estimates efficiency and effectiveness benefits, where USGS defines the former as producing the same output at lower cost and the latter as improving the quality of a current output or producing an output not previously available. Model inputs include measures of the volume of data used in the application, the geographical area included in the data, and the variety of uses of the output, among other things. Model outputs are dollar value estimates of efficiency and effectiveness benefits and the ratio of efficiency benefits to the manual cost of running the application.[15]

We did not find any examples of the USGS GIS benefits estimate model being used within DoD, but some organizations within DoD have conducted a cost-benefit analysis of GIS implementation or of IT-based business process reengineering that includes the use of geospatial data assets. We present two examples here—one for Aberdeen

---

[14] Dickinson and Calkins (1988).

[15] Gillespie (2000).

Proving Ground and another for NAS Patuxent River. Then we discuss conducting a cost-benefit analysis of imagery-sharing cost avoidance. After this section, we discuss how some of the illustrative process and installation savings could be extrapolated across DoD.

Estimating the Savings from GIS Use Across an Installation. The Directorate of Public Works at Aberdeen Proving Ground conducted a cost-benefit study for the implementation of a GIS in 1992 and estimated a net present value of $3 billion in 1992 dollars over an eight-year period. An important part of this estimation was the fact that the data were shared and that a central organization acquired, maintained, and updated the geospatial data, which saved manpower and dollars by not having redundant data acquisition and maintenance costs.

Most of the benefits were in the form of monetized workload reductions. Using various labor rates appropriate for the types of activities or functions listed in Table 7.3, the study estimated the first-year-value of the annual time saved to be in excess of $700,000 in 1992 dollars. In addition, the study estimated first-year savings of $115,000 from not having to repair damaged utilities. The gross value is about $1.3 million in 2005 dollars.

The estimated time saved using GIS is about in line with estimates we collected in the course of our interviews with various personnel performing installation-management-type work. For example, field investigations by architectural and engineering firms were estimated to save nearly 4,100 hours annually in the Aberdeen study. In our study, the Langley AFB Tank Management Program was estimated to save as many as 2,500 hours on field investigations, depending on the amount of construction taking place.

To implement the GIS plan, Aberdeen had estimated costs listed in Table 7.4. The bulk of the expenditures were start-up costs to acquire new geographic data, hardware, and software. Using the Army Materiel Command methodology, the analysts conducting the cost-benefit analysis for the Aberdeen GIS plan calculated a present value (in 1992 dollars) of $1.5 million for project costs and a present value of $4.4 million for project benefits.

Table 7.3
Estimated 1992 Benefits of GIS Implementation by the Directorate of Public Works at Aberdeen Proving Ground

Activity or Function	Annual Hours Spent Without GIS	Annual Hours Spent With GIS	Annual Hours Saved
Master plan basic map information update	667		667
Infrastructure analysis	1,120	280	840
Site approvals	1,680	420	1,260
Utility and base map research			
Transportation planning	160	32	128
Architectural/engineering design	2,560	512	2,048
Tenants research	2,080	416	1,664
Waste water management	425	85	340
Field investigation by architectural/ engineering firms	5,852	1,756	4,096
Map data maintenance	6,247	625	5,622
Facility management	5,760	1,152	4,608
Map research by the utility branch	1,500	150	1,350
Management of facility maintenance contracts	512	56	456
Total	28,563	5,484	23,079

In 2005, a new study of Aberdeen GIS was conducted. It found that additional functions and activities affected by GIS between 1995 and 2005 had saved $8.3 million. In other words, large benefits to GIS implementation had not been anticipated when the original study was conducted. In addition, the 2005 Aberdeen GIS study identified a number of other benefits but acknowledged that they could not easily be quantified. The unanticipated or unquantifiable benefits from the latter study are listed in Table 7.5.

Another study, conducted at the Patuxent River Naval Air Station (Patuxent River), examined the benefits of process reengineering in six core installation management business areas. The process reengineering was made possible by investments in information systems that totaled $13.9 million from 1996 through 1999. The focus of the investment was the creation of the Shore Station Integrated Information System (SSIIS), which migrated existing independent IT systems

Table 7.4
Estimated 1992 Costs of GIS Implementation by the Directorate of Public Works at Aberdeen Proving Ground

	One-Time Cost, $	Annual Cost, $
Cost elements for data acquisition and inputs		
Labor to locate sources of geographic data	720	
Costs for acquisition of new geographic data—contracted	300,000	18,000
Survey control acquisition	60,000	
GIS implementation plan, data dictionary, photogrammetric and planimetric specifications—contracted	40,000	
Training costs	8,000	
Sustaining engineering—contracted	160,000	40,000
Additional information and input of attributes into GIS	120,000	
Subtotal	688,720	58,000
Cost elements for data manipulation and output		
Cost of new computer hardware/software	220,000	
Annual cost for hardware maintenance contract		18,000
Training costs for GIS applications operators	60,000	
Labor and material costs for map generation		18,000
Subtotal	280,000	36,000
Total	968,720	94,000

NOTE: Dollars are 1992 values.

at Patuxent River into a network environment. According to the study, SSIIS provides users with, "a sole point from which to launch other applications, share and edit documents, launch both maps and CADD packages, and operate databases."[16] Simply sharing the data was expected to eliminate redundancies, improve data quality, and ensure data integrity across organizations. Within the six core business areas, several processes involved using and sharing geospatial data assets. The estimated gross

---

[16] *Return on Investment Policies, Concepts, and Methods for Installation Life-Cycle Management* (2002, p. 20).

Table 7.5
Unanticipated or Unquantifiable Benefits of GIS Implementation by the Directorate of Public Works at Aberdeen Proving Ground, 1995–2005

Function	Savings or Cost Avoidance, $
Master plan	7,326,000
BRAC	144,000
Enhanced use leasing	16,000
Utility privatization	43,000
Emergency operation center	581,000
Map production	(a)
Post-operations	(a)
Force protection	108,000
Range operations	(a)
Range sustainment	80,000
Environmental compliance	(a)
Cultural resources	(a)
Natural resources	(a)
Installation restoration	(a)
Military operations	(a)
Total cost savings or avoidance	8,298,000

[a] Not enough information to measure cost savings or avoidance.
NOTE: Dollars are 2005 values.

benefits linked to those processes were about $1.4 million annually, in 2000 dollars (about $1.7 million in 2005 dollars).As with the Aberdeen studies, most of the benefits were calculated as monetized labor savings.

We have attempted to extract the gross benefits of processes using geospatial data assets from the Patuxent River study, but the study itself did not break the benefits into geospatial and nongeospatial benefits, nor did it do so for geospatial and nongeospatial costs. Total project costs and benefits are presented in Table 7.6.

Estimating Cost Avoidance Savings in Sharing Imagery. Cost avoidance is another way to quantify the benefits of sharing geospatial data assets. Our discussion of how installation imagery is shared provides some measures of dollar savings through cost avoidance. We will

Table 7.6
Estimated 2000 Return on Investment for the Business Information Technology Implementation for the Shore Station Management Operations at Patuxent River

	Constant	Present Value
Life-cycle investment	$26.4 million	$22.5 million
Total benefits	$139.1 million	$105.0 million
Net present value		$82.5 million
Cost/benefit ratio		4.7 : 1
Average payback period (years)		0.3

NOTES: Dollars in 2000 values. Projections are through 2008.

also show that avoided costs are easier to calculate in some cases than in others.

In previous sections, we documented the widespread use of imagery from IVT and other sources. Imagery supports many mission areas and yields many benefits, yet it is among the most costly type of geospatial data to acquire and process. The DISDI Office estimated that the one-meter resolution imagery for installations and installation cantonment areas in IVT was worth $3.8 million, and the five-meter resolution imagery for range complexes was worth $450,000. The Army National Guard, which has one-foot resolution imagery for all bases under 60,000 acres and one-meter resolution imagery for all larger bases, has invested about $1 million per year on aerial imagery since 2000, with the imagery being updated every five years.

Given the high price of aerial and satellite imagery, users of geospatial data obviously benefit from sharing imagery whenever possible. The Army's Topographic Engineering Center (TEC), within the Engineer Research and Development Center (ERDC) of the U.S. Army Corps of Engineers, is the U.S. Army's commercial satellite imagery acquisition monitor, with a responsibility to ensure that Army agencies and organizations do not duplicate purchases. The TEC Imagery Office (TIO) "conducts the research, acquisition, archiving, and distribution of current and historical imagery and related products"[17] for customers

---

[17] U.S. Army Corps of Engineers (2005, p. J-1).

such as USACE civil works personnel, installations, and other DoD personnel. TIO is also the Army repository of selected commercial satellite imagery for terrain analysis and water resources worldwide. As such, it supports worldwide military applications and operations, especially the warfighters. TIO works with commercial satellite imagery vendors, NGA, USGS, and other data providers to collect and disseminate commercial satellite imagery and products, aerial photography, and Interferometric Synthetic Aperture Radar/Light Detection and Ranging (IFSAR/LIDAR). As mentioned above, DISDI has provided resources to the TIO since the summer of 2005, including DISDI staff members, so the TIO now also supports the DoD I&E geospatial community to help facilitate more imagery data sharing.

Imagery may be shared under various types of licensing agreements or may be available in the public domain. A common type of license is a restricted user list, which limits the types of users to whom licensees may redistribute data. NGA purchases commercial satellite imagery for itself, DoD, and the Department of Homeland Security, through its ClearView contract vehicle. Users of the ClearView imagery are permitted to redistribute data to state, local, and foreign governments and nongovernmental agencies engaged in joint research projects, but the licensee is obligated to "minimize the sharing of imagery with entities who would otherwise purchase the imagery."[18] Aerial imagery firms often have more open data-sharing policies, allowing organizations who pay to have the imagery firm fly an area freely distribute the imagery to anyone they want.

By acquiring imagery once and sharing often, expenditures can be significantly reduced. NGA's ClearView agreement, by which NGA, TIO, and other DoD agencies gain access to imagery, reduced the federal government's per-unit satellite image costs by roughly 75 percent.[19] Between 2003 and 2005, TIO facilitated the sharing of commercial satellite and aerial imagery valued at approximately $17.5

---

[18] Board on Earth Sciences and Resources (2004).

[19] Board on Earth Sciences and Resources (2004).

million.[20] As one specific example, TIO recently saved Hawaii installations $800,000 by providing one-foot resolution imagery for Oahu. All four Service installations on Hawaii were collaborating to hire a company to fly over the island to create brand new aerial imagery, but through the efforts of TIO, they were able to obtain imagery in the public domain that was only a few months old. The Oahu imagery was shared with each Service. Between the fall of 2005 and spring of 2006, customers included the Air National Guard, the USMC Geo-Fidelis office, the USAF GeoBase office, and the Naval Facilities Engineering Command.

These Hawaii installations were about to pay $800,000 for imagery they eventually got free, which is a clear example of cost avoidance. In many cases, though, the cost avoidance calculation is not so straightforward, as in the case of the IVT imagery. The IVT imagery, which comprised the Space Imaging IKONOS Precision 1-Meter and Space Imaging IRS Reference 5-Meter products, was shared not only with the various BRAC offices but also with certain other agencies that requested access. The DISDI Office tracked those requests through May 2005. If each agency had purchased the imagery anew instead of requesting it from the DISDI Office, the total cost would have been over $23 million. But this is a significant overestimate of the true value of the shared imagery, for two reasons. First, the vendors license imagery knowing that it will be shared with other agencies in DoD, much the same as a software company prices a site license knowing that many people within a single organization will install and use an application. Assuming that each requesting agency would pay $3.8 million for one-meter resolution imagery would be like assuming that a company would buy a site license for each employee. Second, one cannot assume that the agencies requesting the IVT imagery would have purchased the same imagery themselves had they not been able to acquire it free. There were, in fact, alternatives and substitutes for

---

[20] This figure is based on TIO's estimates of the average cost per scene. Separate averages are calculated for each type of sensor. TIO emphasizes that these are rough estimates.

the IVT imagery. One could use the cost of the next best alternative to estimate the opportunity cost.

One alternative to, say, IVT imagery—commercial satellite imagery shared through the ClearView license—or purchasing new aerial imagery is older or lower-resolution imagery available at a much lower price. Publicly available imagery from USGS is available through *The National Map* and even Google Earth. Not all users require the highest-quality imagery or the most timely imagery. Depending on the source (military or commercial) and the technology used for collection (aerial photography; LIDAR; or thermal infrared, electro-optical, synthetic aperture radar, or hyperspectral sensors), imagery resolution can range from less than one foot to 30 meters. The installation IVT imagery had fairly high resolution, but the fact that it was shared with many users does not imply that all users required such high-resolution data. For many applications, imagery may be useful but not essential. If the users have access to recent imagery at low cost they may use it, but otherwise they may rely on nonimage geospatial data or use on-the-ground data collection techniques, such as in-situ monitoring or GPS tracking.

The same principles apply to other types of geospatial data assets that are shared. Surely there is some benefit in the form of cost avoidance—recall the estimate that the federal government's per-unit satellite image costs are reduced by about 75 percent because of NGA's bulk purchases through the ClearView contract—but calculation of the savings can be complicated if substitutes are available or the assets are useful but not essential.

Estimating Effects Across the DoD

Using DoD's *2005 Base Structure Report*, we can conduct interesting extrapolations from the Langley AFB manpower savings from the dig permit and delivery order processes to make ballpark order-of-magnitude estimates of the potential savings for these two processes across all DoD installations in the United States[21] (see Table 7.7).

---

[21] In this estimate, we used only U.S. installations since operations at installations in other parts of the world may have different procedures.

Table 7.7
Order-of-Magnitude Estimation of Potential Annual Savings from Using I&E Geospatial Data Assets for Dig Permitting and Construction Orders at DoD U.S. Installations Based on Estimated Savings at Langley AFB

	Calculation Values
Installation PRV for 2005	$1.535 billion
Total DoD U.S. PRV for 2005	$514.5 billion
Installation PRV as a percentage of total DoD U.S. PRV	0.30
Installation personnel	10,922
Total DoD U.S. personnel	1.995 million
Installation personnel as a percentage of total DoD U.S. personnel	0.55
Average PRV and personnel as a percentage of total U.S. PRV and personnel	0.42
Annual time savings for dig permitting and construction orders at Langley AFB	450 to 2,600 hours per year
Potential annual time savings for dig permitting and construction orders at DoD U.S. installations	100,000 to 600,000 hours per year (equivalent 50 to 300 full-time personnel)

We base the extrapolations on the assumption that total plant replacement value (PRV) and total base personnel (both of which are reported in the Base Structure Report) are rough indicators of the amount of construction work taking place at a typical U.S. installation. The 2005 PRV at Langley AFB is $1.535 billion, or 0.3 percent of total PRV at U.S. installations; Langley AFB had 10,922 personnel in 2005, or 0.5 percent of total personnel at U.S. installations. Averaging those two percentages, Langley AFB represented about 0.4 percent of the total "size" of U.S. installations in 2005. If the 450 to 2,600 annual hours saved at Langley AFB are likewise about 0.4 percent of total savings at U.S. installations, then total savings would have been between about 100,000 and 600,000 hours per year just from using I&E geospatial data assets for the dig permits and delivery orders. Assuming a 2,000-hour work year, the savings equate to between about 50 and 300 full-time personnel.

Note that these estimations are very rough approximations and rely on a very small sample set and the quality of the original estimations provided by the installation staff interviewed. Thus, we suggest that these are, at best, rough estimates, yet they help to convey a sense of how large the potential annual benefits may be. In addition, other common installation processes use I&E geospatial data assets where similar savings could be accrued and estimated from such use. For example, manpower savings could be estimated from using geospatially based NEPA assessment and paperwork processing tools.

Employing the same extrapolation methodology that we used for the Langley AFB examples, we can use PRV and personnel at Aberdeen and Patuxent River to estimate the value of savings from using geospatial data assets throughout U.S. installations. The 2005 PRV at Patuxent River was $2.484 billion, or 0.5 percent of total PRV at U.S. installations; Patuxent River had 9,417 personnel in 2005, or 0.5 percent of total personnel at U.S. installations. Therefore, Patuxent River represented about 0.5 percent of the total "size" of U.S. installations in 2005. If the $1.7 million saved annually at Patuxent River is likewise about 0.4 percent of total savings at U.S. installations, then total savings would have been about $360 million per year.[22]

Similarly, the 2005 PRV at Aberdeen was $3.063 billion, or 0.6 percent of total PRV at U.S. installations; Aberdeen had 12,902 personnel in 2005, or 0.6 percent of total personnel at U.S. installations. Therefore, Aberdeen represented about 0.6 percent of the total "size" of U.S. installations in 2005. If the $1.3 million saved annually at Aberdeen is likewise about 0.6 percent of total savings at U.S. installations, then total savings would have been about $200 million per year (see Table 7.8).[23]

The extrapolations are remarkably close given that the studies differed in timing, methodology, and scope. But one feature that the two studies had in common was that most of the estimated benefits were

---

[22] All dollar values for these extrapolations have been converted to 2005 values. Since this is a rough estimate, we rounded down from $363 million.

[23] Since this is a rough estimate, we rounded down from $209 million.

Table 7.8
Order-of-Magnitude Estimation of Potential Annual Savings from Using I&E Geospatial Data Assets at DoD U.S. Installations Using a PRV-to-Annual-Geospatial-Benefit Ratio

	Aberdeen Proving Ground	NAS Patuxent River
Installation PRV for 2005	$3.1 billion	$2.5 billion
Total DoD U.S. PRV for 2005	$514.5 billion	$514.5 billion
Installation PRV as a percentage of total DoD U.S. PRV	0.60%	0.48%
Installation personnel	12,902	9,417
Total DoD U.S. personnel	1.995 million	1.995 million
Installation personnel as a percentage of total DoD U.S. personnel	0.65%	0.47%
Average PRV and personnel as a percentage of total U.S. PRV and personnel	0.62%	0.48%
Total annual geospatial data asset benefits	$1.3 million	$1.7 million
Potential annual benefits of using I&E geospatial data assets at DoD U.S. installations	$200 million[a]	$360 million[b]

[a] Since this is a rough estimate, we rounded down from $209 million.

[b] Since this is a rough estimate, we rounded down from $363 million.

monetized labor savings, and although in at least the case of the follow-up Aberdeen study certain unquantified benefits were acknowledged, there were no dollar values attached to many of the potential benefits. In addition to the fact that many benefits, such as in increased situational awareness, are difficult to quantify, there are also large, infrequent applications of geospatial data assets, such as the BRAC process, that would not be included in estimates of the day-to-day benefits of using and sharing geospatial data assets throughout the United States.

Conclusion

In this chapter we examined the many effects that result from using and sharing geospatial data assets. Often, the effects are more easily described than quantified or valued. Traditional benefit-cost, ROI, cost-effectiveness, and cost-avoidance analysis can be powerful decisionmaking tools that provide quantitative measures of certain types of effects—mainly efficiency gains, such as time and dollar savings. Although these methods may work well for choosing between current investment alternatives, they are less suitable for planning investment and policy strategies that will lead to positive effects, financial and otherwise, over a longer time period. Furthermore, they are also data- and time-intensive. As discussed above, the Patuxent River ROI study involved researchers literally sitting at workstations with stopwatches to time people performing tasks. Finally, these highly quantitative methods cannot easily reflect important types of benefits. Such methods, although informative, are probably not feasible for DISDI to use by themselves to measure and monitor the effects of DISDI efforts to promote the use and sharing of geospatial data assets.

Our methodology of using together the information flow models, logic models, and benefit-cost analyses, and other quantifying methods (when feasible) provides a more appropriate effect assessment tool. The logic models provide a way to capture the full range of effects. Within the logic model, for selected processes as appropriate, some quantitative methods can be used. We have shown how time and manpower savings can be calculated, but other measures also inform. Noise complaints or map requests, for example, are easily measured. Customers' opinions of the quality and usefulness of outputs from DISDI or Service geospatial offices can be collected.

As discussed in this chapter, the information flow model and the logic model provide different perspectives on effects. The information flow model shows how information is shared and with whom, similar to the information exchange analysis in DoD's Business Enterprise Architecture. The information flow model may therefore support larger efforts within the Business Management Modernization Program within DoD. The information flow model illustrates where DISDI and other organizations fall within the process of sharing geospatial data assets; by contrast, the logic model focuses on a single organization, such as DISDI, and its inputs, activities, outputs, customers, and overall outcomes. Logic models can show policy effects (via changes to inputs and activities), effectiveness effects (such as new or better outputs), and can even show quantifiable efficiency effects, as we demonstrated with the Tank Management Program logic model. Customers in one logic model can have a logic model of their own, as in the case of the Medical JCSG, which was a customer of the IVT program office. By "daisy-chaining" the logic models together, one can get an even better picture of the effects of using and sharing geospatial data assets. DISDI, Service geospatial offices, installations, and other organizations can use our methodology to help understand, assess, and explain the full range of effects from their use and sharing of I&E geospatial data assets.

In the final chapter, we will discuss policy recommendations for DISDI. One recommendation is about maturity models and how they can help guide DISDI efforts to maximize the effect of using and sharing geospatial data assets throughout DoD. Maturity models are not really an assessment methodology and so were not covered in this chapter.

CHAPTER EIGHT

Conclusions and Recommendations for DISDI

As just discussed in Chapter Seven, many missions could enjoy significant benefits from sharing I&E geospatial data assets, both cost savings and in less-quantifiable areas of mission effects, such as improving response time and decisionmaking processes. However, to fully achieve such benefits, the barriers discussed in Chapter Six need to be addressed. Those barriers were grouped into eight categories:

- security concerns and other data restrictions
- different IT system, firewalls, and policies
- lack of communication/collaboration between different functional organizations and disciplines
- lack of knowledge about, interest in, or expertise to use I&E geospatial data assets
- lack of data-sharing policy, standards, and contractual agreements
- data stewards' desire to control access to their data
- lack of on-going high-level program support and investments
- risks from sharing undocumented, poor-quality, and out-of-date data.

DISDI is serving an important role in addressing such barriers. It has already made some progress and could do more. Here, we discuss recommendations for how the DISDI Office can do even more to help the Department of Defense overcome these barriers. These recommendations stem from suggestions by interviewees about what is needed to address the barriers and our analysis of these suggestions, the DISDI Office's current role, and what is appropriate and feasible for the DISDI Office to do.

We have grouped the recommendations into the following categories:

- policy recommendations
- recommendations for coordination and outreach
- recommendations for standards, contracting, and Q/A processes
- DISDI staffing and resource investment recommendations.

Policy Recommendations

The DISDI Office serves an important role setting OSD policy regarding I&E geospatial data assets. DISDI should collaborate with NGA to provide more official OSD policy guidance about the need to share geospatial data assets, about security concerns, and about how to actually share such assets, such as providing guidance about developing MOUs and Memoranda of Agreement (MOAs) for data sharing. Such policy guidance will help DISDI address many of the barriers to data sharing, including security concerns, different IT policies, functional stovepipes, lack of data-sharing policy, data stewards' reluctance to share data, and the lack of on-going high-level program support and investments.

Develop a DoD Instruction About the Importance and Need to Share I&E Geospatial Data Assets

DISDI should, in consultation with NGA, develop official OSD policy guidance, such as a DoD Instruction, stating the importance and need to share I&E geospatial data assets and recommend that organizations facilitate such sharing as much as feasible. The White House OMB Circular, *Coordination of Geographic Information and Related Spatial Data Activities,* A-16 Revised, discusses the need for all federal agencies to coordinate and share geospatial data, and DoD Directive 8320.2, *Data Sharing in a Net-Centric Department of Defense,* discusses the need for coordinating, sharing, and integrating data across DoD. However, no official OSD Directive or Instruction focuses directly on DoD I&E geospatial data assets or the need to share them. Official guidance is needed to help the Services engage more support for their I&E geospatial programs and data sharing; to help installation geospatial staff who are encountering barriers in local IT policies; to help overcome some organizational stovepipe issues; and to help OSD staff and others in DoD understand the importance of these assets and their sharing to the DoD missions. In our interviews, many individuals stated a need for such official OSD policy to help secure and sustain more support from management, IT departments, and other organizations for I&E geospatial data asset development and sharing. As one person stated, "Such 'top cover' would help us work better across different functional departments and with IT staff because we have the OSD instruction to point to." In addition, it can help address the reluctance of some data stewards to share the data by acting as though the data are for their own use only, by helping them realize the broader utility from sharing and the need to share.

Develop OSD Policy Guidance Addressing Security Issues with I&E Geospatial Data Asset Sharing

Currently, there are inconsistencies from installation to installation and Service organization to Service organization about which geospatial data assets are sensitive and require restrictions. There needs to be clear policy guidance about which information is or is not sensitive and DISDI should collaborate with NGA to develop a DoD Instruction providing this policy guidance for some basic I&E geospatial data, such as CIP data. Also, an official process for reviewing other geospatial information for such sensitivities needs to be developed. First, such guidance should clearly state which CIP layers are considered nonsensitive and can be widely distributed, both within and outside DoD, such as installation boundaries, and which data layers may be freely shared within DoD but not as freely outside DoD. Second, the guidance should provide an official process for reviewing potential security risks from sharing geospatial data assets, both individual ones and systems that aggregate and integrate geospatial data from diverse sources. The

FGDC's "Guidelines for Providing Appropriate Access to Geospatial Data in Response to Security Concerns"[1] and the framework for assessing sensitive geospatial data in "Mapping the Risks"[2] both give a good starting point. However, more analysis and procedures would need to be developed to provide appropriate policy guidance for DoD needs.

DISDI should also collaborate with NGA to develop OSD guidance for procedures for sharing sensitive geospatial information both within and outside DoD. For instance, which GIS data layers can be freely shared on the web, shared on .mil with a CAC card, or require other special access, such as through password restrictions. In what cases might some limited I&E geospatial data assets be considered classified? Again, the aggregation and integration issues, such as sensitivities for web-based geospatial data assets that integrate data from many different sources, need to be addressed. When addressing such issues, DISDI should coordinate with the Services, NGA, and other parts of the intelligence community. However, the DoD needs to recognize that there will always be risks, and getting consensus from organizations that are not routinely involved in the tradeoffs between obtaining operational utility from sharing and using data, and those responsible for assessing how the data could be used by adversaries will be difficult. DISDI may be uniquely situated to help DoD arrive at a suitable balance between security and openness.

Develop OSD Policy Guidance About How to Share I&E Geospatial Data Assets

Currently, it can be very time-consuming and manpower-intensive for some installations and other geospatial data asset developers to officially share those, especially outside DoD. They may need to spend time and effort to process approvals through their Service chains of command and to negotiate and write MOUs and MOAs. Currently, different installations and Services develop such MOUs differently. DISDI, in consultation with NGA and the Service GIOs, should develop OSD policy guidance and general procedures about how to share I&E geo-

---

[1]   Federal Geographic Data Committee (2005).

[2]   Baker et al. (2004).

spatial data assets. First, this guidance should clearly state which set of basic I&E geospatial data assets can be broadly shared without requiring an MOU, such as a GIS data layer of installation boundaries or a GIS layer showing installation locations throughout the world. Second, the guidance should include the development of standard MOUs for sharing other I&E geospatial data assets. Such MOUs should include ones for general sharing; for a few specific key mission functions, such as homeland defense and emergency response; and for key organizational types, such as state and local governments and allied governments. These MOUs and other guidance should include reciprocity arrangements that address DoD's need to acquire other organizations' data. Such guidance also needs to be flexible enough to account for unique local conditions and needs in asset sharing, for example, U.S. Air National Guard installations may have different needs in their data-sharing arrangements with their state and local governments from USAREUR ITAM program in sharing data assets with Germany, England, and other allied governments.

Such guidance will help address current barriers, such as IT policies, that make it time-consuming and difficult to share I&E geospatial data assets, especially with organizations outside DoD. Standard MOUs will save time and manpower and make it easier and more attractive for DoD organizations to share these assets.

Recommendations for Coordination and Outreach

The DISDI Office also serves an important role in coordination and outreach regarding I&E geospatial data asset development and sharing within as well as outside DoD. The DISDI Office has already done a lot in this regard, educating DoD agencies about the need to and how to share I&E geospatial data assets. Here, we make some suggestions about how to increase such efforts, which are needed to help spread knowledge and interest in I&E geospatial data assets, to address functional and organizational communication and collaboration barriers, and to help ensure program support and investments.

Continue and Expand on Coordination and Outreach Efforts Inside DoD

DISDI should continue and expand on the work that it has started on coordination and outreach with organizations and individuals across as well as outside DoD. On-going efforts to work closely with the Service GIOs in supporting their program development is an important part of such coordination. Other past and current activities, such as the DISDI Portal, conference presentations, and CoI forums, are all good ways for DISDI to help spread the word about the need and benefits to sharing I&E geospatial data assets. In addition, such activities, especially the CoI and other conferences, have helped to match I&E geospatial data asset developers with new potential users and have helped facilitate more sharing and coordination to more effectively share assets.

The DISDI Office should continue and expand such activities. One way to do this is by helping to facilitate I&E geospatial presentations and preparation of articles by different functional and mission areas, such as at military health and homeland security conferences and in journals. Namely, there is a benefit to reaching outside the traditional geospatial technical community. Participating in the Joint Service Environmental Management 2006 Conference and helping collocated it with the CADD/GIS Geospatial Technologies Symposium 2006 conference was a good start. In addition, DISDI should provide seed funding for travel and writing to knowledgeable installation and other geospatial staff (such as USACE, Service geospatial headquarters offices, and OSD geospatial application developers) who want to engage in such activities.

The DISDI Office could also start an I&E geospatial data asset award program to highlight I&E data asset development, implementation, and sharing, as the OSD environmental office did to help highlight and reward outstanding military environmental activities.

Having DISDI staff work within other DoD organizations, such as NGA and the USACE TEC Imagery Office, should also be continued and expanded on. Such activities are an important way to improve coordination with such organizations. By having staff at the USACE TEC Imagery Office, which helps the I&E geospatial community acquire satellite imagery, DISDI staff are helping the I&E geospatial community learn about and acquire I&E geospatial data assets in a more timely and cost-effective fashion, as was discussed in Chapter Seven.

Assist OSD Organizations in Their Acquisition and Use of I&E Geospatial Data Assets

The DISDI Office should provide more assistance to OSD organizations in their acquisition and use of I&E geospatial data assets. As was discussed in Chapter Six, many OSD organizations that could benefit from using shared I&E geospatial data assets to improve their missions, such as improving their management and resource allocation functions, are not aware of the benefits, are not aware how to acquire the data assets, or lack the technical knowledge to use such assets. With a small amount of assistance from DISDI, such issues can be addressed for a large potential gain to OSD offices.

DISDI has already begun such an activity by having its staff work within other parts of OSD, such as the Office of the Assistant Deputy Under Secretary of Defense (ADUSD) ESOH. The DISDI liaison to the ADUSD ESOH has provided technical assistance and advice on using geospatial information to support the ESOH mission. These activities should also be continued and expanded on, because they are an important way to help other OSD organizations learn about where and how to acquire I&E geospatial data assets, how to use them, and the benefits of using them. Such staff assistance should include both very short and longer-term assignments of DISDI staff to other OSD offices. For example, short-term Intergovernmental Personnel Act (IPA) duty assignments for a month or two at other selected OSD offices (such as the TRICARE Management Agency, the DoD Explosives Safety Board, and the OSD Critical Infrastructure Protection office) would be useful, as well as the longer-term assistance of having DISDI staff as a full-time liaison at the OUSD(I&E) ESOH office for a year.

There could even be a free "rent-a-DISDI staff member" assistance program, where the staff member would spend a few hours at a meeting or review a document to address geospatial technical issues that the OSD office does not have the expertise to handle. The DISDI

Office should provide such limited technical assistance for diverse OSD offices as seems appropriate given the potential benefits to those offices. Another technical assistance task that DISDI staff could provide is to review contract language about geospatial data acquisitions and validate and verify geospatial data quality before OSD acquires or uses such data from other organizations. At one OSD office, geospatial data assets were acquired from an NGO that did not have metadata and concerns arose about data quality.

The DISDI Office should produce high-level geospatial products for improving the decisionmaking of high-level OSD decisionmakers. Such geospatial products could help OSD look at big picture questions. DISDI could help support OSD senior-level decisionmaking through specialized geospatial analysis using existing I&E geospatial data assets. DISDI should be able to quickly produce nationwide maps about OSD bases to answer decisionmakers' questions, such as showing the relationships between military bases and NPS lands when Congress requests information about the relationships. These products need to be quickly developed and customized for high-level decisionmaker needs. The customized BRAC maps using IVT data that were produced for the Office of the Secretary of Defense are good examples of such a product and could be a model for such a process.

Develop an Effective Working Partnership Relationship with NGA
DISDI should continue its efforts to develop an effective working partnership relationship with NGA. Such efforts have included placing a DISDI staff member part time at NGA, developing the MIM installation experiment at Camp Lejeune, and coordinating with NGA on standards developments. From a mission perspective, such close coordination is needed because of the current and potential future use and sharing of I&E geospatial data assets with the warfighting and intelligence communities. NGA has responsibility for DoD's warfighting and intelligence geospatial data asset development and sharing and DISDI has responsibility for the development and sharing of DoD's business geospatial assets, i.e., the I&E geospatial data assets.

DISDI and NGA need to coordinate to develop DoD-wide standards, policies, and procedures, especially when dealing with external organizations, such as the FGDC and state and local governments, to have a single coordinated DoD response. Currently, such coordination is also needed in mission areas where there are joint geospatial responsibilities between the two communities, such as homeland defense and security and training. But, most important, it is needed to ensure that I&E geospatial data assets are shared with the warfighting and intelligence community and vice versa. In some areas, geospatial datasets and tools originally developed for the warfighting and intelligence communities can also be useful to the DoD business domain. For example, the underlying methodology and technological approach from NGA's Palanterra is a geospatial data asset that can be beneficially shared with the DoD I&E geospatial data asset community.

In the spring of 2006, the need for DISDI and NGA to work effectively together became even more apparent. On March 3, 2006, OMB issued a memo requiring that federal agencies, including DoD, designate a senior agency official for geospatial information. This official was to have the role of overseeing, coordinating, and facilitating federal geospatial-related requirements, such as participating on the FGDC steering committee. Effective April 27, 2006, NGA assumed the role as the DoD senior agency for geospatial information management. Because DoDD 5105.6, which was the mission charter for NGA, gave NGA such responsibilities across DoD, NGA was OSD's choice for this interagency responsibility. However, this directive was written in 1996 before DISDI existed and is currently being revised. Given current NGA and DISDI activities, it is worth revising this directive to consider changes in their respective roles and to consider the possibility of having DISDI become the DoD senior agency for geospatial information management in dealing with civil agencies. DISDI's geospatial activities have more in common with other federal agencies than NGA's, and as a policy- and management-oriented agency, not an operational agency that supports the broadest range of DoD geospatial activities, it seems better suited in many respects to dealing with the myriad of problems associated with coordinating DoD geospatial information management activities, and to coordinating with other federal agencies. This latter point is particularly important; because coordination with the other civil federal agencies involves the mission activities and data types that the DISDI community is involved with daily, it makes more sense for that community to take the lead in such coordination.

Even if DISDI does not have the DoD lead responsibility, DISDI should be considered for the role as the official DoD representative in such other federal agency processes as participating on the FGDC committees. That office's expertise and need to coordinate more with other federal agencies in their activities, and because NGA does not have the resources or expertise to represent the DoD business domain in such processes, make DSDI the obvious choice. Whichever organization has the lead, having a single DoD point of contact for coordinating with civil federal agency geospatial activities means that DISDI and NGA need to work closely together for the lead agency to function effectively.

Expand Outreach and Coordination Outside the DoD

Because of increasing demands by nonmilitary organizations for I&E geospatial data assets and the need for coordination with such organizations for many DoD missions, outreach mechanisms also should be expanded outside DoD. DISDI has already begun such outreach activities with its participation in SERPPAS and in the Colorado Project Homeland Pilot. DISDI providing official coordination and outreach with federal, state, and local governments is especially important. DISDI should also, in collaboration with NGA, work more with the civilian federal agencies to coordinate data sharing and data acquisition. For example, DISDI could work with USDA, USGS, and other federal agencies, as well as state and local governments, to help develop deals for sharing the cost of aerial imagery across the United States.

DISDI also should, in collaboration with NGA, take a lead role in trying to acquire data for the U.S. military I&E geospatial data community from other nations and international organizations, such as NATO and the European Union (EU). For example, DISDI could work with NATO to arrange with the EU to acquire some of their data, such as Nature2000, for U.S. installations in Europe. Obviously, DISDI should work with relevant Service organizations, such as USAFE and U.S. Army Europe IMA and ITAM, on such initiatives, to ensure that any such arrangements meet Service needs. Given NGA's extensive relationships with foreign governments, DISDI and NGA should work closely on such international activities.

DISDI should expand on such existing activities and develop appropriate guidance and procedures for sharing data outside DoD, as was discussed in the policy recommendations. In addition, DISDI should provide processes and methods to allow key federal and state organizations and geospatial data clearinghouse/repositories to access basic defense I&E geospatial data assets and to learn where to acquire updated I&E data so that they can update their data versions. For example, DISDI should make sure that the latest installation data boundary layers are routinely provided and accessible in Geospatial One Stop. In addition, the DISDI Portal is currently restricted to .mil and CAC users, which means that most non-DoD organizations cannot access it. DISDI should develop an unrestricted DISDI Portal for nonmilitary users who have a need to access basic I&E geospatial data.

Recommendations for Standards, Contracting, and Q/A Processes

Since standards, contracting, and quality control processes are all key to the sharing of I&E geospatial data assets and not developing and implementing them sufficiently were key barriers identified in Chapter Six, DISDI has an important facilitator role in such processes.

Help Develop and Promote I&E Geospatial Data Standards Development and Adoption

DISDI should, in collaboration with NGA, continue and expand its efforts to help develop and promote the development and use of I&E geospatial data standards. DISDI is already helping to promote common standards across OSD by working with other organizations in the development of OSD geospatial data standards. First, DISDI is working with the OSD Real Property Office to help develop common geospatial definitions for the Real Property Inventory that meet all four Services' needs. Second, DISDI is working with the Services and the CADD/GIS Technology Center to help revise the SDSFIE. Third, DISDI is participating with the NGA Geospatial-Intelligence Standards Working Group to help develop geospatial data standards for sharing across the DoD. For instance, this group is developing the DoD standards for the National System for Geospatial-Intelligence Feature Catalogue Mission Specific Data Standards (MSDS). It is important to note that such standards are added to the DoD Information Technology Standards Registry (DISR) and that DoD elements are required to use standards in DISR. Fourth, DISDI is helping develop data models for federal, state, and local geospatial data sharing for homeland security and emergency response by participating in the Project Homeland Colorado pilot. A key part of this process is DISDI's collaboration with Service GIOs and ensuring that Service needs are met and that their standards and policies are consistent with such OSD efforts. The IVT process for developing the IVT QAP is a useful model for this effort, because DISDI successfully worked with the Services to develop a QA/QC process that met everyone's needs.

These are all fruitful activities that should be continued and will help in the development of useful standards and practices for sharing more I&E geospatial data assets. Such activities need to be expanded to ensure that I&E geospatial data content standards, such as the common naming of installation and environment objectives, features, and attributes, are being developed for different I&E geospatial data assets, which is what DISDI staff members are currently doing with the RPI. Collaboration with NGA and others in the warfighting and intelligence communities is important to develop common data models, symbology, and other standards to facilitate easier sharing with these communities.

In addition, DISDI should help develop official guidance and incentives so that individuals can follow the existing and newly created standards. Unless the standards are implemented throughout DoD, especially by geospatial data stewards who develop and maintain the assets, they are useless. First, DISDI should continue to work with the Service GIOs to ensure that they are providing guidance and incentives to follow the OSD standards. Second, DISDI could provide incentives by providing useful tools that make it easier to follow the standards and useful tools that require following the standards. For instance, if there were a tool that could save an organization time and work hours and that was freely available to DoD organizations and was only usable if the installation geospatial data followed official OSD metadata and data content standards, then installations would have an incentive to follow the standards to use the tool. A DoD geospatially based NEPA paperwork processing tool, dig permitting tool, or AT/FP tool like the installation examples discussed in the appendix are good examples of potential DISDI tools.

Since each Service at the installation level, functional level, or headquarters level has some innovative tools that could be generalized across OSD, DISDI should invest in having these tools become official OSD tools. Not only would these tools help with standards implementation, they would also help facilitate the sharing of geospatial data assets and help increase the number of installations and other organizations benefiting from the use of such tools across DoD, as was discussed in Chapter 7. However, DISDI needs to be sure to coordinate with the Services in the development of such tools to ensure their applicability to all Services.

Provide OSD Policy and Standard Contracting Language for Military Contracts That Involve Digital Geospatial Data and Analysis

DISDI should, in consultation with NGA, provide OSD policy and standards contracting language for military contracts that involve digital geospatial data and analysis. Some DoD organizations, especially ones that lack geospatial technical skills, were writing contracts about geospatial information that did not fully address their own needs. For example, contractors were delivering geospatial datasets without metadata. Contracting language should require that any geospatial data that are delivered must meet OSD I&E geospatial standards, such as requiring FGDC metadata. In addition, the contracting language should require, where economically feasible, that the contract deliver digital I&E geospatial data products, not just hard copy documents and maps, PDFs, or PowerPoint files. In some cases, this may not be economically feasible because of the additional fee the contractor would charge for the digital products. However, in many cases, DoD gives a contractor existing geospatial data, such as GIS datasets, and the contractor adds some value to it, i.e., updates the GIS datasets slightly, in which case there should be only a limited extra fee. DISDI should help review OSD contracts that involve geospatial information and ensure that OSD offices know that DISDI can provide such assistance. The recommended policy guidance about the importance of sharing I&E geospatial data assets should include a section discussing contracting needs.

Ensure That Quality I&E Geospatial Data Are Made Available for Sharing and Are Shared

QA is the process put in place to ensure that during a digital geospatial data creation project, the data meet established quality standards. QC involves the monitoring of project results to make sure that they comply with the quality standards, such as statistically spot-checking data. In creating geospatial datasets, most developers try to meet certain quality standards depending on the purpose of the data and level of accuracy needed. However, as discussed in Chapter Six, we found that many I&E geospatial data asset developers do not invest sufficient time and money in developing thorough QA/QC procedures, especially when they assume that only their own organization will use the data. Having appropriate data-quality procedures is very important for I&E geospatial data asset sharing. For example, the IVT development process had an extensive QAP to ensure that the highest-quality data available were included and that users understood that quality. All IVT data had to meet BRAC auditing standards, which also contributed to the extensive quality assurance process. Each data layer had an official data steward who signed off on its quality and each installation commander also had to sign off on the data provided for the installation. Each installation was required to provide explanations if some data were not provided. We found that I&E geospatial data assets developers and users, especially within OSD, valued the product more highly because of this extensive QA/QC process.

Using the IVT experience as a model, DISDI should, in consultation with NGA, provide OSD guidance on standard QA/QC process to develop official authenticated DoD I&E geospatial data. In addition, the DISDI Office should provide updated official authenticated DoD data, like the IVT data, on a regular basis through the DISDI Portal and to OSD offices and others who have a need to use such data. Since I&E geospatial data change over time, it is important to update these data, which were created in 2003. In fact, DISDI should work with the Services to develop an official OSD CIP that is updated at least every year. The IVT data and process provide a model and foundation for the development of such an OSD CIP. The IVT data layers should be expanded on to add additional data, such as official installation environmental, Real Property Inventory, and training data layers.

DISDI Staffing and Resource Investment Recommendations

The tasks mentioned above represent a large workload for the current DISDI staff. DISDI currently has one director, four contracted staff members, and some limited funds to allocate for projects. Given such tight resources, it is important that DISDI is managed wisely. We present three suggestions for how to do this. The first has to do with possible ways to augment DISDI's staff. The second provides DISDI a way to better assess past and on-going investments. And the third suggests a way to manage DISDI's investments.

Examine the Benefits from and the Feasibility of Temporarily Expanding the Number of DISDI Staff

DISDI does not have enough staff members to perform the many diverse functions recommended in this monograph. In fact, it is impressive how much they have already been able to accomplish with so few people and such a short amount of time. Given the potential benefits from having more DISDI staff members to perform the recommendations suggested here and the potential cost savings to DoD as a whole with more investment in DISDI staff, we suggest exploring the idea of increasing the number of DISDI staff. Given the potential mission effects and savings, adding another staff member or two could more than pay for itself in the long run. In addition, since the program is still fairly new, extra staff members are needed to develop the program over the next year or two. Because of the current DoD financial situation, it may be difficult to acquire staff positions or additional contractor support. Another useful way to augment DISDI staff is to have knowledgeable geospatial staff work part time at DISDI as IPAs from other DoD organizations. Such staff could come from Service headquarters, major or functional commands, installations, and other OSD organizations, such as NGA. Having such IPAs would also help with coordination and information sharing with the DoD organizations that the staff come from. DISDI could also have IPAs come from other federal agencies, such as geospatial data and application experts from USGS or even EPA.

Use the Information Flow and Logic Model Methodology to Help Assess Effects

DISDI should use the methodology we developed, as discussed in Chapter Seven, to help assess the past and on-going success of its investments.

Specifically, we recommend applying a methodology that consists of three elements:

1. an information flow model to understand the range of organizations using and sharing an I&E geospatial data asset
2. a set of logic models to map out how the inputs, activities, and outputs of an organization's data development, use, and sharing lead to outcomes for different customers
3. to the extent possible, when the data are available, employment of a variety of methods for quantifying the logic models.

The DISDI Office, as well as the Service GIOs and other organizations, can use this approach to help understand, assess, and explain the full range of effects from the development, use, and sharing of I&E geospatial data assets. Such assessments will help the DISDI staff understand which of their investments and activities have had the most benefit and where and how. DISDI staff can use the results of such assessments and understanding to help plan future investments. For example, DISDI staff could apply the methodology to the SERPPAS initiative and DISDI's role in it, to better understand its effect on which customers and how to best develop and invest in similar initiatives with state and local governments in the future. In addition, applying this methodology will also enable the DISDI Office to more effectively explain to its senior managers the full range of benefits and effects of its activities across the GIG.

We also recommend that the DISDI Office, the Service GIOs, or functional mission organizations apply this methodology to individual mission areas, such as training or military health, to assess the effect on the area. This study only scratched the surface of analyzing the effect on individual mission areas, as demonstrated by the discussions throughout this monograph. A useful follow-on study to this work would be to take what was learned in this study about a given mission area and expand on it to assess the effect on the mission areas. For example, a more in-depth assessment of the effect on training could be conducted by applying our methodology to the OSD and Service sustainable range programs and by interviewing more training range operators, planners, and developers about how they use I&E geospatial data assets and the effects on their missions. Such an assessment would help show more specific mission benefits in a mission area and could be used to help facilitate more support for I&E geospatial asset development, use, and sharing from that mission area commander and other DoD senior managers.

Establish Processes for Managing Future Investments by Applying the GAO Maturity Model

Our recommendations focus on processes, policies, and organizational relationships that will enable DoD to maximize its return on investment in geospatial data assets. Long-term improvements in processes, policies, and organizational relationships can be planned systematically using the IT Investment Management (ITIM) maturity model developed by the GAO. In our review of the application of this model, we found that when applied appropriately and in the right circumstances, as GAO found, this maturity model can provide three key capabilities to organizations that adopt it:

1. a rigorous, standardized tool for internal and external evaluations of an agency's IT investment management process
2. a consistent and comprehensible mechanism for reporting the results of these assessments to agency executives, Congress, and other interested parties
3. a road map that agencies can use to improve their investment management processes.[3]

The ITIM model has five stages of maturity, with all but the first stage having distinct critical processes and key practices. As an agency adopts the key practices, it moves to higher levels of maturity in managing IT investments. There are three types of key practices: organizational commitments, prerequisites, and activities. Table 8.1 shows the five stages of the ITIM maturity model and their corresponding critical processes.[4]

Table 8.1
The GAO ITIM Stages of Maturity with Critical Processes

Maturity Stages	Critical Processes
Stage 5: leveraging IT for strategic outcomes	Optimizing the investment process Using IT to drive strategic business change
Stage 4: improving the investment process	Improving the portfolio's performance Managing the succession of information systems
Stage 3: developing a complete investment portfolio	Defining the portfolio criteria Creating the portfolio Evaluating the portfolio Conducting post-implementation reviews
Stage 2: building the investment foundation	Instituting the investment board Meeting business needs Selecting an investment Providing investment oversight Capturing investment information
Stage 1: creating investment awareness	IT spending without disciplined investment processes

---

[3] U.S. General Accounting Office (2004, p. 5).

[4] U.S. General Accounting Office (2004) has much more detail on each maturity stage, critical process, and key practice.

GAO reports that most agencies it has evaluated are currently operating at Stage 2. We would argue that this is where DoD is with respect to its investments in I&E geospatial data assets. Although an agency operates at one stage, it may have initiatives under way that will lead it to a higher stage of maturity. For example, DISDI is currently working on various processes to develop an investment portfolio, which is the third stage of the ITIM maturity model.

Review of I&E geospatial data asset investments may fall to recently created Investment Review Boards (IRBs) as part of DoD's Business Management Modernization Program. By law, IRBs will review all business system modernization investments exceeding $1 million and assess the business mission effects as they relate to supporting warfighting requirements and improving financial accountability. As discussed in Chapter Seven, many effects cannot be easily quantified or measured, but DISDI will have to provide some justification for investments in I&E geospatial data assets. Therefore, the maturity model framework can be a powerful tool for demonstrating how new investments and initiatives by DISDI will lead to a more mature process of managing I&E geospatial data assets.

We recommend that the DISDI Office apply the GAO's ITIM maturity model to help manage investments. Doing so can create both internally and externally focused benefits for DISDI. As a framework for planning future activities and initiatives, the maturity model can help provide long-term organizational goals. The framework can also serve as a basis for resource and investment justification to external agencies and groups, such as the IRBs. The methodology is endorsed by GAO. It establishes a road map to improved performance of the geospatial data asset portfolio, and it provides an alternative to traditional benefit-cost and related analyses that tend to be static and cannot easily measure many benefits.

Conclusions

Clearly, installation I&E geospatial data assets are being shared with many different organizations in many different ways inside and outside the DoD. The assets enable many diverse mission areas from the installation to the OSD level. The effects from such use and sharing relate to both efficiency, such as cost, manpower, and other time savings, and to effectiveness, such as improving operations and decisionmaking, as well as to secondary benefits such as improving communications and working relationships. Besides significant cost and manpower savings, these mission effects include significant mission benefits such as facilitating more time on a training range and improving training operations, enabling more rapid deployment and improved warfighting logistics, allowing faster incident response, and improving the quality of life and facilities investments for military communities. However, the use of I&E geospatial data assets in many of these areas has just begun and more needs to be done before DoD can fully enjoy the benefits. Trends are that such data sharing and benefits will likely increase and reach even more users through more web-based portal-sharing approaches. However, barriers to increased data sharing still exist and include security concerns, lack of on-going high-level program support and investments, lack of data-sharing policy and standards implementation, and functional stovepipe issues. The DISDI Office and the Service geospatial information offices serve important roles in addressing barriers to data sharing and in facilitating more I&E geospatial asset development and sharing across DoD.

It is difficult to assess and quantify the full mission effect of using and sharing I&E geospatial data assets. Nonetheless, we have developed a three-phase methodology for such an analysis. Step 1 is to develop an information flow model of the data sharing to understand which organizations are affected. Step 2 is to apply logic modeling to map inputs to activities, outputs, customers, and the full range of outcomes. The logic model is a useful tool for assessing the effect of using and sharing geospatial data assets. Step 3 is to use analytical techniques, such as cost-benefit analysis, to quantify parts of the logic models where manpower and other cost effects can be quantified. However, it is important to include specific effects that cannot be quantified, such as improving operations, planning, and decisionmaking; improving knowledge, communications, and working relationships; and policy and process changes. This methodology gives the DISDI Office and different levels within each Service a tool to use to help demonstrate and document the full range of effects of using I&E geospatial data assets.

By implementing methodology such as the one described here and the types of policy recommendations outlined for the DISDI Office, the development of I&E geospatial data assets and sharing will continue to increase and to accrue significant financial and operational benefits across DoD, helping to save tax dollars and lives and improve mission performance.

Details on How I&E Geospatial Data Assets Enable Business-Related Missions

During our research, we found that I&E geospatial data assets enable many diverse missions throughout DoD in many different ways. As discussed in Chapters Two and Three, any mission function that involved some type of location information has the potential to use geospatial information to help track, manage, view, or analyze that information. This appendix describes in detail the diverse applications in the 12 traditional installation mission uses that employ I&E geospatial data assets, mainly business-related mission uses:

- base planning, management, and operations
- emergency planning, response, and recovery
- environmental management
- homeland defense, homeland security, and critical infrastructure protection
- military health
- morale, recreation, and welfare: enhancing quality of life
- production of installation maps
- public affairs/outreach
- safety and security
- strategic basing
- training and education of the U.S. military
- transportation.

For each mission area, examples are provided to show how I&E geospatial data assets support the mission. The examples were chosen to illustrate how many different organizations and people use these assets in many different ways to support diverse mission activities. Note that the mission areas where I&E geospatial data assets support traditional warfighting operations were discussed in detail in Chapter Four, so they are not discussed here.

Here, we present examples of how I&E geospatial data assets support the mission for each of these 12 mostly business-related mission areas, since I&E geospatial data assets are and can be used in many different ways within a given mission area. Over 130 examples are discussed in this appendix. To not overwhelm the reader by presenting all these examples, we recommend choosing to read about selected mission areas of most interest.

To avoid some detail, these examples focus mostly on how the assets support the mission rather than going into details about which data elements are shared and how. (This topic was covered in Chapter Two.)

Note that the mission areas in our examples are not mutually exclusive; some I&E geospatial data asset uses support multiple missions at the same time. Therefore, some examples could be discussed in multiple places. We discuss them in detail only once and refer to other parts of the appendix where they could also be discussed.

Base Planning, Operations and Management

Many geospatial data assets, such as GIS datasets about roads and buildings, were developed to help manage and run military installations. Base planning, operations, and management support includes installation planning; construction and sustainment of real property, such as the maintenance and repair of existing facilities; and the development of the Real Property Inventory and providing support to military operations at the installation. In supporting such functions, the use of geospatial data assets have helped improve the efficiency of the management and operational installation processes by integrating diverse datasets and automating formerly manual processes. Mission effects often include significant savings in man-hours, as formerly mostly manual processes are automated with digital geospatial data applications. Such savings were discussed in Chapter Seven. However, the mission effects also include saving costs, improving the management or analysis process, and even helping to reduce worker frustration.

In examining the many examples identified in our study and discussed below, it seemed natural to group these examples into three main mission area subcategories:

- facility and infrastructure planning and construction
- facility and asset management
- base operating force support activities.

Facility and Infrastructure Planning and Construction

This subcategory refers to any type of site planning and construction activity, including buildings, roads, utilities, and other facilities assets. For example, at Naval Air Station Whiting Field, Florida, installation GIS data were used to help the installation determine where to locate a new air traffic control tower. At Marine Corps Air Ground Combat Center at Twentynine Palms, California, GIS analysis helped determine the best placement of wind energy turbines. Geospatial data are used to help plan, locate, and analyze the construction of such facility assets. They are also used to help assess and manage base construction contracts and track installation capital improvements.

One of the most common application examples in this mission area was using GIS data to help assess whether a location is suitable for construction given a wide range of building considerations. Such GIS analysis saves time and money and often improves the decision-making process because it is easier to integrate and view effects when siting the facility asset. For example, planners at APG use the installation GIS system to help make faster and more effective decisions about where to build new buildings and testing facilities. The GIS system quickly shows constraints on the land being considered for development. Constraints include, but are not limited to, flood plains, wetlands, Bay Critical Areas, bald eagle nest locations, quantity distance arcs, airfield clear zones, utility access, aquifers, training areas, surface danger zones, unexploded ordance (UXO) locations, and cultural/archaeological areas. A GIS analysis was conducted to determine where to place a new high-speed track for automotive testing. It showed the relationship with wetlands, wood duck mating area, airfields, etc. Twenty to 30 decisionmakers, including planning, airfield, environmental, engineering, USACE, and commander staff met and reviewed four scenarios in the GIS to choose the best location. The GIS staff had created the four scenarios in two days, where in the past, without GIS, one scenario would have taken three weeks to create.

To help streamline this siting process, some installations have developed specialized application tools. For example, Ramstein AB has developed an "Environmental Planning Analysis Tool" to look at environmental constraints in planning activities, such as building construction plans. The user loads this application on the desktop and it runs off the data on the server using the CIP and environmental databases. If someone wants to build something on the base, they draw the structure in its proposed location into the system to conduct a constraints analysis. The system provides environmental constraints, including information about wetlands, endangered plants and other species, and underground storage tanks. The user sees a map showing what constraints may restrict building at that location.

Military headquarters organizations also use I&E geospatial data assets to help with construction planning and operations. For example, ACSIM has supported the Army's transformation master planning of the Senior Review Group of Army G-3 Transformation by supplying map products showing planned construction at different Army installations around the United States that are receiving new missions as a result of the 2005 BRAC decisions.

I&E geospatial data assets are also used to help with contractor oversight as construction is carried out, such as in the USAF Air Mobility Command, which uses its "AMC GeoBase Console" (the USAF AMC GeoBase MapViewer) to help manage base construction contracts across multiple installations.

Facility and Asset Management

Installation geospatial data assets are used extensively to help track, assess, manage, renovate, maintain, and repair installation facility assets, including the grounds, buildings, infrastructure, and other facility assets. Given the advancements in information and geospatial technologies, such as GPS, and the ability to link into installation databases, such as Service work order relational databases, geospatial data are used to manage some very detailed assets and processes. Such assets include a wide range of assets external to buildings, such as manhole covers, fire hydrants, flagpoles, dumpsters, and storm water drains, and assets internal to buildings, such as floor space, equipment, and furniture. Diverse processes are supported, including workspace management, grounds maintenance, facility use, management of tenants' utility bills, fire hydrant maintenance, dumpster tracking, snow removal, and dig permitting processes. Basically, the I&E geospatial data assets help with managing the contracts, operations, and costs associated with facility services and asset and building space assessment and management.

A common maintenance use of I&E geospatial data assets is to help more efficiently and effectively assess and conduct repair and maintenance of buildings, roads, and other facility assets and infrastructure. For example, Marine Corps Base Camp Pendleton, California, uses I&E geospatial data assets to help with the maintenance and repair of sewer lines, water supply lines, and other utilities. The installation geospatial staff members give technicians a digital map showing the locations of the water and sewer systems. The technicians use it with a GPS receiver to locate water valves, sewer line segments, and other system items needing repair or maintenance work. The use of the GPS and GIS-generated maps has made it a lot easier to locate such assets in the field. Currently, the installation is videotaping the sewage lines and will link each photo to its corresponding segment in the GIS database so that technicians can see photos as well.

Another type of facility maintenance issue has to do with snow removal. Dover AFB, Delaware, has developed a GIS-based Snow Removal Tool to track snow removal in near-real-time. After a snowplow clears a paved surface areas, such as a road, runway, or taxiway, the operator enters that information into an interactive web-based tool. This tool helps to prioritize snow removal and shows which areas are back in service. Information from the field is more quickly and widely made available about which transportation routes and infrastructure are available for use.

Installation geospatial information has also been used to help develop contracts and manage contractors that provide services to help run installations. The creation of maps for key contract support, such as mowing, helps clarify the wording of the contracts and ensures that both parties understand what is required. It also is used to provide contract oversight. For example, at Langley AFB, staff members used their GIS system to calculate the acreage cut by grass cutters before paying them—a more accurate method than the old way of field inspection, which has improved the accuracy and timeliness during this process and has saved money on grass-mowing contracts.

Installation geospatial data assets are also used to help assess where to place future investments; for example, commanders at APG have used a GIS system to help prioritize facility investments. Since the GIS system links directly into detailed base databases, such as the Integrated Facility System (IFS) and Installation Status Report (ISR), they can look at all the buildings that have roof problems. These buildings show up red on the GIS map, which helps commanders see the problem areas and decide on priorities for roof repair investments.

Another common use of geospatial data assets is to streamline dig permitting processes. Before anyone can dig into the ground at any U.S. military installation, someone must make sure that the digging equipment will not damage anything that may be buried underground, such as a water pipeline. The approval of various installation organizations must be obtained before digging is allowed at that specific location. Keesler AFB has implemented an automated Dig Permit Tool that is web-enabled and integrates mapping functionality. The tool reduced the time to approve a permit from weeks or days to hours. Customers log on to the GeoBase web page, fill in the form, and click on the Submit button; simultaneous email notifications go out to all reviewing agencies. During the submission process, the proponent can mark on a map the dig location with an interactive Red Line Tool. The map will then identify every utility within the marked area. After each reviewer completes a review, the system forwards the request to the final approval office. When the permit is approved, an email notification is sent to the customer.

Benefits include time savings, customer satisfaction, and ease of initiating a dig permit. Not having to hand-carry a dig permit all around the base for approval represents more than just time savings: The reduction in frustration factor is another major benefit. After the dig permit is ready for approval, having a "one stop" point of contact is also beneficial. Another benefit is that reviewers have a map to look at, showing the utilities possibly affected by the dig. This process helps avoid costly mistakes of accidentally breaking utility lines.

Base Operating Force Support Activities

Installation geospatial data assets are also used for base planning for operating force support activities, improving the efficiency and effectiveness of managing such assets. Operating force support includes air operations support (airfield operations and aviation support), port operations, and other types of operations support.

A common use of geospatial assets is to support airfield operations development and use. For example, USAF AMC used installation imagery and other geospatial data to conduct aircraft parking analyses, Camp Butler used such data to help with runway planning. and NAS Patuxent River conducted a GIS analysis to determine where to build new aircraft hangers for the Joint Strike Fighter.

The Navy used geospatial data assets to help with port management, development, and operations. For example, the U.S. Navy at Pearl Harbor, Hawaii, uses GIS data to help plan ship berthing, namely, looking at which ship to bring into which port. In the future, it plans to link utility data into this system, since utility use can be high aboard ship, and to calculate the best way to allocate utility assets. Navy Region Northwest uses its RSIMS to support port operations. It created a 3-D model of a wharf to help evaluate the spatial relationships between the wharf and specific Navy vessels. For example, a 3-D model of a Nimitz class carrier was positioned next to the wharf model to see if there was sufficient space to permit the docking of this ship.

Emergency Planning, Mitigation, Response, and Recovery

Emergency planning, mitigation, response, and recovery include any application of I&E geospatial data assets for the planning, mitigation, response, and recovery to natural disasters and other emergencies. I&E geospatial data assets support activities at the installation level—such as responding to a disaster at a base—and at a regional, national or international level—such as supporting military response to Hurricanes Katrina and Rita. They also support issues such as the reconstitution of the base and evacuations of both military and civilian personnel. Improving operations, planning, knowledge, communications, working relationships, and decisionmaking are all key mission effects of using and sharing geospatial data assets for emergency planning, mitigation, response, and recovery. Faster and better decisions during an emergency response can save lives.

This mission area overlaps others, such as homeland defense/critical infrastructure protection, antiterrorism/force protection, and environmental management. Homeland defense examples are discussed below. Here, we focus on more traditional emergency response roles, such as preparing for any type of disaster, responding to natural disasters, and emergency 911 calls.

Examples are organized into these categories:

- emergency planning, mitigation, and preparedness
- emergency response
- emergency recovery.

There are overlaps, too, between these areas, especially the first two. For example, many I&E geospatial tools are used for both emergency planning and response.

Emergency Planning, Mitigation and Preparedness

I&E geospatial data assets are used by defense emergency response planners to prepare, mitigate, and plan for all sorts of manmade and natural disasters, including earthquakes, hurricanes, floods, typhoons, terrorist attacks, and other homeland defense incidents. Such activities include analysis and simulations of potential incident scenarios, analysis of emergency operation and evacuation plans, development of sophisticated emergency response systems, and training exercises.

For instance, Kunsan Air Base, on the southwestern coast of the South Korean peninsula, has used 3-D GIS data to help prepare for the typhoon season, and Travis AFB in California has used GIS to support earthquake simulation training exercises. At Travis AFB, the use of I&E geospatial data assets has improved the communication and coordination process as well as the efficiency of the training exercise. The wing commander and key emergency personnel all see the same map of the incident in near-real-time; in the former manual process, installation staff depended on hand-carried sketch pads or whiteboards with information from phone calls. Similarly, Camp Butler staff members have used the base 3-D GIS model to create simulations for planning, analysis, and emergency response exercises to tsunamis.

Providing the common installation/operational picture for situation awareness during a training exercise is a common use of I&E geospatial data assets. The Navy Dahlgren site in Virginia uses installation GIS data to provide a Common Installation Picture for emergency response exercise that tests chemical sensing robots. This activity supports R&D testing of sensing technology and robot performance and enhances emergency response preparation.

Another common use is to help in developing and documenting emergency operations plans, and even to help make such documents dynamic. For instance, at NAS Patuxent River, GIS analysis supports the development of the Emergency Operations Plan (EOP), the Oil and Hazardous Substance Spill Contingency Plan, and the Fuel Operations Plan. For oil and hazardous material spills, analysts have a set of maps showing locations of fuel transfer sites, fire mains, storm water drainage, and manholes; these maps provide input to all these plans. In addition, analysts have made these plans dynamic by linking them directly into the live web-based installation geospatial data.

Another interesting example shows how the development of a spill response GIS system can support emergency planning, preparation, response, and even recovery, and how a regional tool can be developed. Environmental staff members at Naval Support Activity (NAVSUPPACT) Naples, Italy, are using GIS to help develop Navy site spill response plans to improve response and control. They have integrated detailed geospatial information at several Navy sites, including imagery; base boundaries; drainage information; hazardous waste facilities; petroleum, oil, and lubricants (POL) storage areas; and spill control equipment locations with other site-specific spill response information into a customized GIS application. Users can now easily access and analyze information about spill response, such as where the spill is going and where is the closest equipment to help contain the spill. Since different bases have different concerns and data, the tool is customized for different site conditions. So far, analysts have developed this tool for two sites: Naval Support Activity Capodochino, Italy, and NSA Support Site Gricignano, Italy. They are developing an on-water version of the tool to address the spill response needs of ships at ports. This tool version will include oil dispersion modeling using the General NOAA Oil Modeling Environment (GNOME) model under varying conditions to make better informed decisions. They also hope to expand the application to the entire region and have started developing versions for other sites in the area.

The advantages of their GIS approach is that it provides easy access to site-specific information needed to manage a spill, aids in planning and staging control equipment in relation to threat, helps with quick inventory of assets when a spill occurs, and helps fire and disaster preparation staff perform their mission. In addition, the system is a communication tool at command and control meetings, can help improve response time for communicating command decisions, provides a means to display complex information, provides detailed information on response capabilities and response staging areas, and provides analysis tools to devise spill-prevention procedures. For post-spill analysis, the system can map the progress of the spill and cleanup, can be used to visualize the extent of both surface or subsurface contamination, and can be used for site characterization postincident to help with recovery.

Emergency Response

Emergency response use of I&E geospatial data assets includes a range of applications—from sophisticated customized emergency response tools to supporting Emergency Operations Center and 911 systems, using web systems, GIS tools, data, and map books.

Many geospatially based emergency response tools have been developed to help plan and coordinate faster and more effective responses to any type of emergency. Such application tools are web-based as well as stand-alone desktop systems for use in the command post. Often, both are developed to prepare for large-scale emergencies when the web is not available. For instance, Ramstein AB, Germany, has a strong program to prepare for and support emergency response exercises and contingencies using GIS-based tools. Ramstein GIO staff members have GIS data and applications on a laptop to support actual emergencies and exercises. This system is called the Emergency Response Tool Suite. The system provides commanders with situational awareness. Staff members respond to and support the battle staff in the event of an emergency or emergency response exercise. They do this using a portable laptop or PC hooked into the base Intranet network, but they can use a stand-alone system if the network is down. The data for the incident are entered in real time. As calls come in giving such information as incident location and attribute changes, GIO staff members change the map accordingly. If need be, they can perform real-time plume analysis of chemical spills to help plan quick evacuations and response.

Headquarters organizations also use geospatial information to help prepare and respond to disasters. In preparation for tracking Hurricane Rita, USAF headquarters operations branch produced a map with the projected path, terminal airspace, and air traffic control as the hurricane was approaching the Gulf Coast. Analysts updated the map every few hours as the hurricane approached. NOAA had a live web link with the latest information, so whenever staff members there viewed their map, they had the latest tracking data in it.

During the emergency, staff with the Air Force Office of the Deputy Chief of Staff for Installations and Logistics needed to know the Gulf Coast infrastructure, such as pipelines, DoD depots, etc. USAF headquarters operations branch produced a situational awareness map with key locations on it, such as the DoD depots, Katrina area of responsibility (AOR), and oil for emergency use. When the military was called up to respond to Rita, this mapping system helped with the initial deployment decisions. It also was used to help answer questions about potential effects on oil infrastructure.

Many installations use GIS for their 911 emergency response system and installation Emergency Operations Center, such as at APG. At APG, emergency services use the base GIS map for their 911 system and during emergencies, such as Hurricane Isabel, and emergency training exercises, base GIS staff members have supplied real-time support to the EOC. The ability to quickly show overflight photos and checkpoints, to plot traffic control points, to model air plumes, and to measure distances "has proven vital in emergency operations."[1] In early 2006, the Navy also started implementing a Navy-wide GIS-based dispatching system for E911 and other installation emergencies, called the Naval Emergency Response Management System (NERMS). Namely, fire department, emergency medical services, and police department dispatching and support will be supplied through NERMS, which uses installation GIS data of roads, building addresses, etc.

Sharing geospatial information and collaborating with other organizations, especially federal, state, and local agencies, is important for this mission. For example, during a large brush fire that was burning partly on Vandenberg AFB, California, Vandenberg AFB staff flying in helicopters used GPS to locate the fire boundary lines and combined the information with vegetation burn history in their GIS. They shared this information with CDF. Together, they used it to calculate where the fire would burn next, thus improving joint firefighting efforts.

Emergency Recovery

Installation geospatial data assets are also very useful for supporting emergency recovery planning, analyses, management, and operations. Such support can include activities ranging from assessing environ-

---

[1] Aberdeen Proving Ground (2005).

mental changes to determining the effect on personnel, facilities, and infrastructure. For instance, as discussed in Chapter Three, Camp Butler's 3-D GIS modeling system was used to assess and develop ecological recovery measures after an accidental training fire to help prevent erosion and storm water runoff problems. Another environmental restoration example is that after Hurricane Isabel, staff members at NAS Patuxent River added shoreline changes to their GIS and these data were used to help with the shoreline restoration.

Helping to assess and conduct restoration activities for installation buildings, facility assets, and infrastructure is another common use of I&E geospatial data assets at an installation, a region, or a functional command. For example, at NAS Pensacola, installation GIS data were used to help assess the restoration and demolition needs of historic buildings damaged during Hurricane Ivan. Hurricane Ivan caused extensive damage to Naval Air Station Pensacola— a National Historic Landmark covering 82 acres—and its environs on September 15, 2004. This site includes the remains of the 19th-century Pensacola Navy Yard, the nation's first Naval Air Station and first Navy pilot training center. Maps were developed to show the extent of the damage to these historic sites and were used to help examine restoration and demolition options and also to communicate the status of the facilities and the proposed plans to the public and to relevant federal, state, and local agencies, such as the National Park Service, the Florida State Historic Preservation Officer, and the Advisory Council on Historic Preservation.[2]

A potential personnel application is the use of GIS to examine the change in military population demographics near a military medical center after a major disaster. For example, USAF Keesler Medical Center in Biloxi, Mississippi, was supporting 35,000 retirees before Hurricane Katrina hit. By early 2006, it was still unclear how many of these retirees would return and how many would decide to live else-

---

[2] The Advisory Council on Historic Preservation (ACHP) is an independent federal agency that promotes the preservation, enhancement, and productive use of our nation's historic resources and advises the President and Congress on national historic preservation policy.

where. GIS analysis could be used by staff to examine the new expected population demographics to help assess the workload and future status of Keesler Medical Center.

Environmental Management

I&E geospatial data assets are used to support numerous environmental management functions. In fact, we identified over 100 examples of the use of these assets to support environmental management activities within DoD. I&E geospatial data assets are used to help monitor, research, assess, and manage a wide range of environmental issues, including anything that has to do with air, water, land, and other natural resources, as well as cultural resources, chemical wastes, and noise concerns. I&E geospatial data assets are used to help with compliance and official reporting for numerous environmental regulations and programs, including the Clean Air Act (CAA), the Clean Water Act (CWA), the Resource Conservation and Recovery Act (RCRA), NEPA, and the Endangered Species Act (ESA). For example, NAS Patuxent River has developed and used a GIS-based "NEPA Checklist" application tool to document and help speed up the NEPA process and to help inform users about NEPA requirements.

I&E geospatial data assets play an important role in ensuring that our installations have clean air; safe drinking water; healthy watersheds and ecosystems; and proper chemical, hazardous materials, solid waste, and hazardous waste management and disposal. They also are used to help maintain the land and ecosystems, so that the U.S. military can have realistic and diverse training environments. Using I&E geospatial data assets for strategic issues, such as sustainability and helping to manage an installation's entire environmental management system, are good examples that can have a significant effect on management and installation processes. I&E geospatial data assets also play an important role in installation environmental research and development activities, such as with OSD Strategic Environmental Research and Development Program (SERDP) projects. For example, in the SERDP Ecosystem Management Project at Fort Benning, university scientists and other researchers conducted field monitoring of plants and environmental conditions. They entered this information into a GIS and analyzed it with installation GIS data to help advance installation ecosystem understanding.

In this section, we present examples of how I&E geospatial data assets support diverse environmental functions. These examples are organized into categories based on the natural groupings for the range of applications that we identified. They are organized as follows:

- cleanup and management of toxics and hazardous materials, spills, and wastes
- cultural resources
- cnvironmental impact and strategic assessments and analyses
- noise management
- species, natural resource, and land-use management
- water management.

Another category—environmental education and outreach—is discussed below in the public affairs and outreach section. Since environmental issues are so interrelated with other mission areas, other environmental examples are also discussed in other sections of this appendix, such as soil erosion issues in the training section and interrelated operational safety and environmental concerns in the safety and security section.

Cleanup and Management of Toxics and Hazardous Materials, Spills, and Wastes

I&E geospatial data assets are used to help monitor, track, manage, and clean up environmental solid and hazardous wastes, and at restoration projects. They also are used to help manage toxics and hazardous materials, spills, and wastes.

Installations and the Services use I&E data assets extensively to help with hazardous waste management and cleanup. Specialized geospatially oriented tools have been developed and are used to track and monitor environmental conditions, comply with environmental laws, and conduct public outreach. For example, Camp Navajo, New

Mexico, has an Environmental Data Management System (EDMS) web site for managing and mapping shared site data for environmental cleanup of UXO and toxic chemicals. This Army National Guard base is being cleaned up to state and federal standards according to the Resource Conservation and Recovery Act under the DoD's Installation Restoration Program (IRP). The EDMS web site was developed to give users access to the environmental data generated under the Camp Navajo Installation Restoration and Open Burn/Open Detonation Area Closure Program. Environmental data presented on this web site include chemistry, geologic, geographic, hydrogeologic, and unexploded ordnance data from 1980 to present. Users can interactively display information about restoration sites. This web site has helped with documentation, communication, and public outreach.

Diverse I&E geospatial data assets are used to help with hazardous waste cleanup and other objectives at the same time. For instance, the U.S. Navy invested $400 million in clearing UXO, debris, bomb fragments, and scrap metal from Kaho'olawe, a Hawaiian island that was used extensively as a naval gunfire and aerial bombardment target. GPS, GIS, and a spatial database are all being used to help safely clear the area as well as to map, document, and protect more than 3,000 archaeological finds. The island will be protected for cultural and educational activities and archaeological investigations. The GIS information will be able to help support these activities as well.[3]

An example of using I&E geospatial data assets to help with toxics management can be seen with asbestos. Asbestos consists of a group of fibrous minerals that were historically used extensively in the manufacture of more than 3,000 products, including building materials, insulation, and brake linings. Its use continued to increase through the 1970s. At that time the evidence against asbestos as a health hazard was made clear and the EPA and Occupational Safety and Health Administration (OSHA) began to regulate asbestos. Exposure to asbestos fibers can cause the development of certain forms of lung cancer, such as mesothelioma. Asbestos is not always an immediate hazard.

---

[3] For more information see: "Cleaning up Kaho'olawe: U.S. Navy Completes Massive UXO Project" (2004, pp. 10–11).

Only when materials containing asbestos are disturbed or those materials become damaged does it becomes a hazard. When asbestos-containing materials become damaged, the fibers separate and may then become airborne creating a hazard to human health. MCAS Cherry Point, North Carolina, has a web-based asbestos database management system that shows where asbestos is known to be on the installation. The system ranks exposure risks since the last inspection. It uses the installation CADD floor plans, which show asbestos locations colored in red, yellow, and green, where red identifies areas where asbestos is exposed and a risk, yellow represents areas of possible risk, and green shows areas that are sealed off and are not currently a risk. This system is used to assess where to invest dollars for asbestos cleanup and remediation. For example, a door had broken which exposed asbestos since the last inspection. This Asbestos Management System alerted management that precautions and immediate cleanup were needed.

Cultural Resources

Cultural resource management addresses cemeteries, archaeological sites, and historic buildings.[4] U.S. military installations must follow U.S. regulations and laws to help document and preserve these cultural resources. Regulatory requirements for cultural resource management are found in the National Historic Preservation Act, the American Indian Religious Freedom Act, the Archaeological Resources Protection Act, and the Native American Graves Protection and Repatriation Act. I&E geospatial data assets support the identification, tracking, assessment, management, maintenance, protection, and information sharing and public outreach about these archaeological and historic assets and sites.

Many installations enter archeological location and attribute information into their installation GIS data system to help track, manage, and protect them. For example, Naval Air Weapons Station (NAWS), China Lake, California, covers over 1.1 million acres in the upper Mojave Desert. The installation contains extensive archaeological and

---

[4] Historic buildings include older buildings and those of the Cold War period, from 1946 to 1989.

cultural areas of concern, such as Native American pictographs. Many of these pictographs are of such quality and worldwide cultural value that the installation conducts guided tours for archaeologists, other scientists, and the public. These assets are tracked in an installation GIS database, which is used to help manage, record, and preserve these resources while accomplishing the installation's missions. At NAWS China Lake as at many other installations, such data are also consulted and assessed when siting an activity at the installation, whether a new building or institution of a training exercise, to make sure that these archaeological treasures will not be harmed during the activity.

Installations also use geospatial data assets to record, track, and manage historically significant buildings. Many military installations include historic buildings and landmarks that are protected by historic preservation laws. I&E geospatial assets are used extensively to help with the upgrading and maintenance of such facilities, especially when renovations or new construction is being planned. Considerations need to be made for preserving the historically significant and protected aspects of these sites. In some cases, even visual effects need to be considered. For example, the U.S. Military Academy, at West Point, New York, has used 3-D modeling to help assess the potential visual effects of new building construction because the site is on the national historical register, is a national historical landmark, and is in an official National Heritage area (the Hudson River Valley). When considering the placement of a new stadium, the West Point planners needed to consider the potential visual effect to both the military installation and the surrounding community. Using CADD and GIS technologies, they built a 3-D model of the proposed structure in location with surrounding buildings and vegetation and used a line-of-sight tool to assess visual corridors both on and off the installation. Visual considerations included the aesthetic, architectural, and visual effects, as well as potential light pollution, because of concerns from the neighboring community about stadium lights.

Besides historic buildings, I&E geospatial assets are also used to help with management and outreach about historically significant cemeteries. For example, at APG, GIS supports information management and outreach relating to historic cemeteries. There are 14 cemeteries on the post, some of which are in active training areas. GIS staff used GPS to identify their locations and have linked in some photos of headstones and relevant documents. Historians and families can use this information without having to visit the site, since many of these sites are not readily accessible. In addition, sometimes base staff escort family members to the actual sites if they are in active testing areas. NPS staff from Arlington National Cemetery visited Aberdeen to look at this GIS system because managers there are thinking of using GIS in a similar way to help visitors assess and learn about grave sites. Sharing information about this application with the NPS is an example of how a military I&E geospatial application has benefited another federal agency, the NPS.

Environmental Impact and Strategic Assessments and Analyses

One of the most important environmental uses of I&E geospatial data assets is to help conduct strategic analyses and impact assessments. Environmental impact assessments involve both formal and informal analyses of how some sort of military activity, whether testing equipment, training exercise, or building a new building on an installation, affects the environment, including effects on species of concern, habitat, ecosystems, water and air quality, and wetlands. Strategic environmental assessments range from analyzing future habitat trends to assessing future encroachment to developing more sustainable installations. Such assessments often occur at the installation level but also can occur across multiple installations by different function commands and regions and even at OSD. We present diverse examples here to illustrate this point.

A common use of I&E geospatial data assets is to analyze the environmental impact of an installation activity including installation development projects, training, and weapons testing. Such analyses support formal legal processes, such as NEPA requirements, and nonlegal assessments, such as looking at an activity's effect on natural resources. For example, in 2004, Fort Benning built a new major training range, called the Digital Multi-Purpose Range Complex. Installation GIS data, such as information about habitat and species of concern and wetlands locations, were used to assess the environmental impact of the construction and use of this training range. The GIS maps that were generated were also used in the NEPA process and official NEPA documents. The assets also support the environmental assessments for the redesign of ranges and testing areas (such as briefly mentioned in the training section of Table 7.2), where Camp Lejeune staff used GIS analysis to help reconfigure a range away from the river to minimize any environmental impact on the river. I&E geospatial data were used to help assess and minimize the effects on T&ES and wetlands and to help with the official NEPA assessment process, as well as to supply maps for the NEPA documents and public outreach process.

Many installations even develop special automated tools to help assess the environmental impact from construction activities, such as Ramstein AB's Environmental Planning Analysis Tool, which was discussed above because of its dual mission use for environmental and base planning and management.

Another type of environmental analysis that uses I&E geospatial data assets is strategic analyses of environmental concerns, such as predicting changes in land use and habitat over time. Because of environmental laws and pressures from encroachment, all larger installations perform environmental impact analyses and strategic assessments to analyze effects from current and future activities both on and off the installation. A common strategic environmental assessment conducted by installations, as well as Service and other DoD environmental staff, is encroachment analyses. I&E geospatial data assets are combined with geospatial information about areas surrounding the installation and used to help assess and address installation encroachment. Encroachment can be defined as outside pressures for resources, airspace, waterways, and land that affect or have the potential to affect military training, testing, and readiness. Encroachment issues include endangered species habitat on military installations, competition for radio frequency spectrum, protected marine resources, competition for airspace, air pollution, noise pollution, and urban growth around military installations.

All the Services and installations develop plans and partnerships to "buffer" testing and training lands and other installation operations from the effects of encroachment, such as the ACUB projects discussed in Chapter Three. A buffer project involves different mechanisms to ensure that off-base open spaces and habitat are protected that benefit installations, such as when private landowners volunteer to donate conservation easements.[5]

I&E geospatial data assets are used to help identify, prioritize, and choose buffer areas, such as with the USMC encroachment partnering program. Working with the community partners is an important part of this geospatial analysis. The USMC process begins with the establishment of a regional conservation forum, consisting of installation staff, local and regional community planners, conservation groups, and other nongovernment stakeholders. The conservation forum develops a conservation plan with regional goals and maps key areas of mutual interest and opportunity. For example, at both Camp Lejeune and MCAS Beaufort, GIS data were key in helping to map, assess, develop, and choose conservation buffer project areas in collaboration with local conservation forums. MCA Beaufort also used the GIS maps to help in public outreach about their MCAS Beaufort Encroachment Partnering Initiative. Similarly, NAS Fallon in Fallon, Nevada, has used I&E geospatial data assets to help in the installation's encroachment partnering program with Churchill County.

To help compare encroachment factors across different installations, the U.S. Army Environmental Center has developed a GIS-based Encroachment Condition Module (ECM) to geospatial calculate the effects of encroachment across Army bases. A consistent ECM score is calculated at each base regarding each encroachment factor (wetlands, cultural resources, T&ES, noise, air quality regulations, etc.) and the effects on training (such as no dig, no smoke, no live fire, and no night training) by intersecting polygons and their attributes.

---

[5] "A conservation easement is a deed restriction landowners voluntarily place on their property to protect resources, such as productive agricultural land, ground and surface water, wildlife habitat, historic sites or scenic views. They are usually by landowners ("grantors") to authorize a qualified conservation organization or public agency ("grantee") to monitor and enforce the restrictions set forth in the agreement. Conservation easements are flexible documents tailored to each property and the needs of individual landowners. They may cover an entire parcel or portions of a property. The landowner usually works with the prospective grantee to decide which activities should be limited, to protect specific resources." American Farmland Trust (1998).

The user can click on each polygon to see the effect on training by time of day and year. By March 2006, the ECM had been applied to Fort Pickett, Virginia.

I&E geospatial data assets are also being used to help strategically plan for sustainability. Over half a dozen Army installations have begun developing sustainable installation plans that develop and address long-term sustainability goals related to the interrelationships of mission, the environment, and community. Such installations include Fort Lewis, Washington; Fort Bragg, North Carolina; Fort Carson, Colorado; Fort Benning, Georgia; Fort Stewart, Georgia; Fort Campbell, Kentucky; and Fort Jackson, South Carolina. GIS data play a critical role in the development of such plans. In fact, in their sustainable installation analysis and development process, Fort Jackson staff members have identified many different types of GIS data, such as land-use and species location and trend information, as key data needs for this process.

DoD regional environmental assessments also use I&E geospatial data assets and combine them with other federal, state, and local data. OSD's use of I&E geospatial assets, as well as civilian government data, to help in strategic environmental assessments was seen with the Southeast Regional Partnership for Planning and Sustainability as discussed in Chapter Three. A Service example is found with the USMC Western Region Environmental Plans Office, which uses I&E and other geospatial data assets to help assess what other federal agencies, states, developers, and utilities are doing off base that might affect USMC installations in the Western region. For example, analysts at the base assess whether a significant housing development is planned within a flight corridor and where DOE and the State of California are planning to put in a new energy corridor and what the effect might be on nearby USMC bases.

Noise Management

Because of more and more urban and suburban encroachment around installations that used to be in isolated rural areas, noise complaints from training exercises, whether air or ground, have increased significantly over the last 10–25 years. Installations use I&E geospatial data assets to help analyze, manage, minimize noise effects and complaints, and to conduct outreach. Often, such analyses and management involve the use of noninstallation geospatial data, such as census population data, to help locate nearby populations in relationship to installation training activities.

In 1998, the NAS Patuxent River facility received complaints from community members regarding UAV operations over the Northern Neck of Virginia. Although the station worked with the public during the process of writing an Environmental Impact Statement for operations at the Patuxent River Complex, citizens expressed concern about UAV overflights, which they claimed were extremely noisy, circled the same area for extended time periods, and were unsafe.

In 1999, the station used GIS to determine alternative UAV routes in an effort to reduce the effects on the Northern Neck of Virginia community. Population densities were mapped and routes were reassigned based on these densities and other operational considerations, such as safety zone areas. Since the new UAV routes were established, the station has received no UAV noise complaints from the Northern Neck of Virginia residents. The use of I&E geospatial data assets helped reduce noise complaints and risk, i.e., improved operational safety by not flying over as many populated areas.

Installations also develop tools to help assess and track noise complaints, such as at MCAS Cherry Point, North Carolina. Staff members there used I&E geospatial data assets to create a noise complaint tracker. They geocoded county addresses and then assessed the relationship with their training activities, including fire, smoke, and aviation. The system helps to determine which are valid complaints and to improve community relations.

I&E geospatial data assets have also supported noise management activities at regional and headquarters levels (as discussed in Chapter Three with the "United States Air Force and National Park Service Western Pacific Regional Sourcebook" initiative). The assets are also used in models to estimate the noise from ground training. One such model, the USACE Small Arms Range Noise Assessment Model (SARNAM), estimates the noise from small arms fire at ranges, determining exposure level values for both humans and animals. SARNAM has been used for this purpose by multiple installations within the Army and USMC.

Species, Natural Resource, and Land-Use Management

I&E geospatial data assets are also used to help monitor, track, manage, and research species, habitat, and ecosystems, especially concerns about T&ES. They are also used to help manage other natural resource issues, including hunting, fishing, and forest and timber management. I&E assets are used extensively to help with land-use assessments and management, such as predicting, assessing, and mitigating soil erosion from military training. (Erosion issues are discussed below in the training section; here, we give examples relating only to species, forest, and hunting management).

I&E geospatial data assets are used extensively to help manage and assess the status of T&ES[6] and their habitat as well as threats to the habitat and ecosystem integrity. All four Services use the assets to help research, track, monitor, assess, and manage T&ES and other species of concern throughout the United States. Such species include green sea turtles and Gulf sturgeon at Eglin AFB, Florida; wolves at Camp Ripley, Minnesota; desert tortoises at Fort Irwin, California; the California gnatcatcher at MCAS Miramar, California; and *Abutilon menziesii* (an endangered plant) at Pearl Harbor Naval Complex, Hawaii. For example, at APG, GIS analysis was used to identify the cause of a bald eagle death and to help develop measures to prevent the deaths of other eagles. GIS staff assessed the cause of the death of a bald eagle, first by using GPS to determine its location, then by examining the location in relation to live firing at a training area, power lines, and other hazards. The assessment showed that the death was

---

[6] The federal Endangered Species Act is the law that protects species and can restrict federal activities that affect them. The ESA's purposes "are to provide a means whereby ecosystems upon which endangered species and threatened species depend may be conserved, [and] to provide a program for the conservation of such endangered and threatened species" (16 USC §1531b, or Farley and Belfit, 2001). To accomplish this objective, the U.S. Fish and Wildlife Service (USFWS) establishes a list of species in danger of extinction, identifies the habitat needed for conservation, and develops plans to recover the species, and listed species are protected from being "taken" without express authorization of the USFWS.

caused by the bird flying into the power lines. As a result of this incident, an additional analysis was performed to help assess where to place reflectors on power lines to prevent further eagle deaths.

I&E geospatial data assets are used to help research, monitor, track, and fight "invasive species," which are a growing environmental concern as they threaten military installations. An invasive species is one that is nonnative (or alien) to the ecosystem under consideration and whose introduction causes or is likely to cause economic or environmental harm or harm to human health.[7] Plants, animals, and other organisms (e.g., microbes) can all be invasive species. Invasive species cost the United States billions of dollars each year through their effect on agriculture, forestry, and public health. They also can affect military readiness.[8]

At Fort Carson, Colorado, environmental staff plan to use remote sensing data to monitor, track, and help fight the spread of tamarisk—an invasive species that is hurting the water supply and riparian habitat at the Teller Reservoir at the south border of the installation. Tamarisk is an invasive tree species that threatens many water sources in the Southwest because it has an extensive spreading root system, con-

---

[7]   Executive Order 13112, "Invasive Species," February 3, 1999.

[8]   The military, as well as U.S. civil agencies, are very much aware of potentially significant effects to civil society and military activities from invasive species because of the experience with the brown treesnake on Guam. Shortly after World War II, the brown treesnake was accidentally transported from its native range in the South Pacific to Guam, probably as a stowaway in military ship cargo. As a result of abnormally abundant prey resources on Guam and the absence of natural predators and other population controls, brown treesnake populations reached unprecedented numbers, with densities as high as 12,000 per square mile. Snakes have caused the extirpation of most of the native forest vertebrate species; thousands of power outages affecting private, commercial, and military activities; widespread loss of domestic birds and pets; and considerable emotional trauma to residents and visitors alike when snakes invaded human habitats and posed threats to small children. Approximately every third day there is a snake-caused power outage somewhere on Guam with costs estimated at $1 million to $4 million each year from direct damages and lost productivity. Effects on the U.S. military include time delays and potential dispersal of brown treesnakes to other places in routine military traffic, the need to employ special practices in military training in the Western Pacific Region, and the need for additional procedures for the management of wildlife on military lands on Guam. For more information see U.S. Department of Agriculture (n.d.); National Biological Information Infrastructure (n.d.); and Westbrook and Ramos (2005).

sumes an enormous amount of fresh water, and can threaten riparian habitats.

I&E geospatial data assets are also used to help with forest and timber management. They are used in various ways, from tracking and managing forest stands to fire modeling for planning controlled burns. For example, the Naval Support Activity (NSA) Crane, Indiana, uses GIS to help manage timber and other forest resources throughout this 62,000-acre installation. The goals of the program are to preserve and protect forests, restore and maintain species diversity, and produce high-quality hardwood sawtimber. Forest information, such as the location of stands that need harvesting, and species information, such as the locations of the federally endangered Indiana bat, are input to a GIS and used to help manage these natural resources. In fact, GIS is used to track unusually large, high-quality white oak trees that are the official wood used for the restoration of the USS *Constitution*.

Planning controlled burns to enhance forest and ecosystem health is another common use of I&E geospatial data assets. For example, Eglin AFB uses remote sensing and other GIS information to help determine where to conduct controlled burns for managing the longleaf pine forests. The longleaf pine habitat is key to the long-term survival of the endangered RCW and other species of concern and is dependent on fires. Eglin AFB staff members have developed a series of GIS-based models to assess ecosystem dynamics and relationships. One of these models is a GIS-based controlled-burn prioritization model. This model has been used by managers at Eglin AFB to help with a variety of base decisions related to land use, such as where and when to conduct military operations and where and when to have controlled burns. This planning activity allows Eglin to sustain base training and other military missions while protecting endangered species and the ecosystem. This GIS-base model also provides a good example of cross-service sharing, because Camp Lejeune staff has taken this model and adapted it for their own installation needs.

Many installations also use I&E geospatial data assets to help assess, manage, and operate their hunting and fishing programs. For example, MCAS Cherry Point, North Carolina, uses the installation GIS data to help manage and implement its hunting program. Small game, such as turkey and other fowl, and large game, such as bear and deer, can be hunted during hunting season by Service members, their families, military retirees, and installation staff who have base access. The installation GIS system is used to assess any potential conflict between training and hunting areas by showing when and where training and hunting locations would overlap. If there is a potential conflict, the trainer notifies the installation game warden so that hunting is not allowed during the training exercises.

Water Management

I&E geospatial data assets are used to help track, manage, and assess water issues related to drinking, surface, and ground water. They have supported installation-level clean drinking water programs, storm water and watershed management, and the analysis of water pollution problems. For example, at NAS Patuxent River, the installation geospatial data system is used to help update the drinking water plan. The Navy also used installation GIS data to perform underground 3-D plume analysis to examine ground water pollution at NAS Jacksonville, Florida, and Naval Weapon Station Charleston, South Carolina.

A good application of geospatial data assets for storm water and watershed management is at USMC Camp Butler in Okinawa, Japan. As discussed in Chapter Three, the environmental management staff at Camp Butler created a detailed 3-D map of the drainage on and around Camp Butler, which was used to address water runoff issues from an accidental training fire. Besides helping with managing such water runoff concerns, this system is also used to help with flood management, storm water infrastructure investments, tsunami simulations, spill response, and environmental education. Many of these applications were discussed in other parts of this monograph. Here, we present two different Camp Butler examples showing how this 3-D model supported, at the same time, water management concerns and other mission areas, emergency and force protection planning, and response. The environmental management staff used the 3-D GIS model of the drainage system to identify low points and placement of new storm water drains to prevent flooding. The environmental management staff also developed a GIS-based predictive spill model. They used it to examine surface flows for spill planning to help prevent water contamination and the spread of hazardous or toxic contaminants, such as fuel. The tool is designed to help emergency responders, to assist with tabletop spill drills, to help identify areas that need protection, and to integrate this information into the base spill contingency plan. The model was used in a force protection exercise at MCAS Futenma. In the exercise, it was assumed that a 7,500-gallon jet fuel tank (JP-8) was hit by a rocket propelled grenade (RPG), resulting in a fuel spill. Analysts animated the spill using the GIS data to show where it would flow both on and off the base, which helped improve communications and exercise effectiveness. The base commander, military police, and fire department staff all liked the GIS-based animation, since it was easier to understand than the topographic maps usually used for such a force protection tabletop drill.

Even nonmilitary organizations have used installation geospatial information to help with watershed and ground water management. For example, USDA researchers have developed a GIS-based model to quantify riparian vegetation groundwater use in the San Pedro River Basin in southeastern Arizona and northern Mexico. Many believe that the presence of large-scale groundwater pumping in the nearby urban areas of Sierra Vista and Fort Huachuca has created a cone of depression, which has, or will soon, diminish the base flows in the river. The GIS-based tool is an accounting model that merges a vegetation map, including parts of Fort Huachuca, with component vegetation groundwater use models to help management agencies determine the total riparian vegetation groundwater use in the San Pedro Basin and how this use would change with different management strategies, such as prescribed burns.

Homeland Defense, Homeland Security, and Critical Infrastructure Protection

In the last few years, I&E geospatial data assets have been used to help plan, prepare, and analyze homeland defense and homeland security activities and for critical infrastructure protection. Before discussing such support, it is important to define each of these terms. According to DoD Joint Publication 3-26, *Homeland Security*, homeland defense is defined as

> The protection of United States sovereignty, territory, domestic population, and critical infrastructure against external threats and aggression or other threats as directed by the President. The Department of Defense is responsible for homeland defense. Homeland defense includes missions such as domestic air defense. The Department recognizes that threats planned or inspired by "external" actors may materialize internally. The reference to "external threats" does not limit where or how attacks could be planned and executed. The Department is prepared to conduct homeland defense missions whenever the President, exercising his constitutional authority as Commander in Chief, authorizes military actions.[9]

Homeland security, as defined in the National Strategy for Homeland Security, is a concerted national effort to prevent terrorist attacks within the United States and to minimize the damage and recover from attacks that do occur. The Department of Defense contributes to homeland security through its military missions overseas, homeland defense, and support to civil authorities."[10]

Critical infrastructure protection (CIP) includes "actions taken to prevent, remediate, or mitigate the risks resulting from vulnerabilities of critical infrastructure assets. Depending on the risk, these actions could include: changes in tactics, techniques, or procedures; adding redundancy; selection of another asset; isolation or hardening; guarding, etc." Such actions can also include planning, analysis, and management to support the protection and operation of infrastructure.

The specific definition of critical infrastructures has evolved over the years. Often the terms "national critical infrastructure" and "key assets" are used to include both infrastructure as well as other assets

---

[9] Department of Defense (2005b, p. GL-9).

[10] Department of Defense (2005b, p. I-3).

that are important to the nation. The Department of Defense defines national critical infrastructure and key assets as "the infrastructure and assets vital to a nation's security, governance, public health and safety, economy, and public confidence. They include telecommunications, electrical power systems, gas and oil distribution and storage, water supply systems, banking and finance, transportation, emergency services, industrial assets, information systems, and continuity of government operations."

On December 17, 2003, President Bush issued Homeland Security Presidential Directive 7 (HSPD-7) clarifying executive agency responsibilities for identifying, prioritizing, and protecting critical infrastructure. This directive defined critical infrastructure slightly differently, as seen in Table A.1. This table also shows lead agencies. The diverse lead agencies shows how DoD needs to collaborate with the other federal

Table A.1
Critical Infrastructures and Lead Agencies Under HSPD-7

Lead Agency	Critical Infrastructure
Department of Homeland Security	Information technology Telecommunications Chemicals Transportation systems, including mass transit, aviation, maritime, ground/surface, and rail and pipeline systems Emergency services Postal and shipping services
Department of Agriculture	Agriculture, food (meat, poultry, egg products)
Department of Health and Human Services	Public health, health care, and food (other than meat, poultry, egg products)
EPA	Drinking water and wastewater treatment systems
Department of Energy	Energy, including the production refining, storage, and distribution of oil and gas, and electric power (except for commercial nuclear power facilities)
Department of the Treasury	Banking and finance
Department of the Interior	National monuments and icons
Department of Defense	Defense industrial base

SOURCE: HSPD-7.

agencies in critical infrastructure protection and how DoD's lead role is to protect the defense industrial base. The defense industrial base means "the Department of Defense, government, and private sector worldwide industrial complex with capabilities to perform research and development, design, produce, and maintain military weapon systems, subsystems, components, or parts to meet military requirements."[11]

For this mission area, DoD also collaborates and shares I&E geospatial data with private sector companies, such as utility companies. In fact, HSPD-7 requires that DHS and other federal agencies collaborate with "appropriate private sector entities" in sharing information and protecting critical infrastructure (Par. 25).

Many installations, regions, Service headquarter offices, and major commands are preparing for homeland defense, homeland security, and CIP every day in their security and emergency response planning, development, and operations. I&E geospatial data assets help improve the efficiency end effectiveness of assessment and planning activities and improve decisionmaking, incident response times, and collaboration. Homeland-defense-related security concerns, such as antiterrorism and force protection, and emergency response measures, such as general emergency response tools, are discussed in the emergency response and safety and security sections of this appendix. Other organizations within and outside DoD also use I&E geospatial data assets to help plan for homeland defense and homeland security. In fact, because of the need to collaborate with other federal, state, and local agencies, such as DHS and EPA and jurisdictions near the installation, I&E geospatial data assets are being shared more and more with nondefense organizations to support this mission. For example, the U.S. Coast Guard has created an Enterprise GIS for distributing core GIS data and functionality to its personnel in its missions relating to marine safety, port security, and law enforcement on U.S. waters. It includes GIS datasets about Navy ports from NAVFAC, as well as other military installation information. The sharing and use of military installation data in this Coast Guard system help support the homeland defense and critical infrastructure missions.

---

[11] Department of Defense (2005b, p. GL-7).

For discussion purposes, we divide examples into two categories:

- critical infrastructure protection assessments
- homeland defense and homeland security planning and preparation.

Critical Infrastructure Protection Assessments

Different defense organizations use I&E geospatial data assets to help map, assess, and address the critical infrastructure protection conditions, vulnerabilities, and investment needs at military installations, as well as in other parts of the United States.

Installations map and assess their own critical infrastructure vulnerabilities using geospatial data assets and use this information to help address such vulnerabilities. For example, NAS Patuxent River GIS staff mapped critical assets, including water systems, electrical substations, and fuel farms and examined populations living near them. They created detailed maps of the fuel farms and evacuation routes in case of an emergency. Often, such data are placed on a classified system, such as at Marine Corps Base Camp Pendleton, California. At this installation, geospatial staff have identified, as specified by security forces, critical infrastructure for the installation, including critical nodes and vulnerabilities. These data were combined with unclassified I&E geospatial data, such as roads and buildings, and was placed on a classified system accessible to security forces and the commander of the installation.

Different DoD organizations use I&E geospatial data assets to help protect individual critical infrastructure types, such as water and power, at U.S. installations across the world. For instance, the USA-CHPPM has developed Emergency Response Plans (ERPs) for U.S. Army installations, primarily because of the Bioterrorism Act of 2002. This act requires that water utilities conduct vulnerability assessments (VAs) of their water systems and to use their findings to update their ERPs. ERPs also help the installation prepare and respond to biohazard emergencies and provide key information about the water infrastructure, such as water customers, and an overview of the water system.

GIS is used to create maps for the ERPs and to help assess the water system and emergency response plan maps for use in the ERPs. These maps are also used in tabletop ERP planning exercises with installation staff.

OSD also conducts critical infrastructure vulnerability assessments for defense installations. For example, as discussed in Chapter 7, the Office of the Assistant Secretary of Defense for Homeland Defense CIP conducts critical infrastructure protection mission area analyses at installations to assess vulnerabilities in assets that support diverse missions, and the office wants to use I&E geospatial data assets to save time and money in such efforts.

I&E geospatial data assets are even used to help assess nonmilitary critical infrastructure issues when defense activities could be affected. USAF headquarters operations branch in charge of air spaces and ranges assessed critical infrastructure overflight concerns about nuclear sites. After the September 11, 2001, terrorist attacks, the DoD and FAA asked, what if we imposed commercial and military no-fly zones over all high-level nuclear sites in the United States? A GIS analysis by USAF headquarters operations branch showed that such a policy would have shut down all commercial and military air traffic over the East Coast and was not viable or implemented.

Homeland Defense and Homeland Security Planning and Preparation

Installations and the Services' regions, functional commands, and headquarters also use I&E geospatial data assets to help prepare and plan for homeland defense, as discussed elsewhere in this appendix. Besides military installations and other Service organizations' homeland defense planning and preparation uses of I&E geospatial data assets, other defense and nondefense organizations also use such assets for homeland defense and homeland security. We discuss those uses here.

Within DoD, NGA's Office of the Americas has a key role in supplying geospatial information for homeland defense and homeland security, and it has developed tools to help. NGA has developed Palanterra, a web-based spatially enabled, real-time common operational picture (COP) of information to describe, assess, and depict physical features and geographically referenced activities referred to collectively as geospatial intelligence (GEOINT). NGA's Palanterra is serving as a common framework and foundation for web-based Geospatial Intelligence decision support and analysis, visualization, and dissemination of homeland security and critical information protection information for the United States. Palanterra uses I&E geospatial data assets.

Another interesting collaborative homeland security activity that used I&E geospatial data assets is Global Mirror. Global Mirror, a DHS- and FEMA-supported emergency preparedness exercise, was conducted on May 10–12, 2004, in Colorado Springs, Colorado. The exercise used GIS to support a WMD scenario. The exercise also crossed several jurisdictions: Peterson Air Force Base, the City of Colorado Springs, and El Paso County, Colorado. Fifty-two federal, state, and local agencies and other organizations participated in the exercise. It used geospatial data integrated from Peterson AFB, the City of Colorado Springs, El Paso County, Colorado Springs Utilities, USGS, NGA, and FEMA to provide a common operating picture for situational awareness. Geospatial technologies were also used to integrate information on City of Colorado Springs and Peterson AFB emergency response vehicle locations that was provided by an automated vehicle location (AVL) system and to integrate the visualization of hazardous release of materials events using the Defense Threat Reduction Agency's (DTRA's) Hazard Prediction Assessment Capability (HPAC).

Even state agencies use military I&E geospatial assets to help with homeland security. As discussed in the main text, both Colorado and Pennsylvania are creating Homeland Security/Public Safety geospatial web-based portal systems for data sharing among federal, state, and local government agencies and first responders. Similarly, the Maryland Emergency Geographic Information Network (MEGIN) is a coordinated information portal to serve geospatial information to the emergency management community at all levels of Maryland government in the event of a homeland security or other type of emergency. MEGIN provides and integrates relevant information, presents the emergency in GIS-generated maps, offers automatic, controlled access to data, and allows backup to a secure off-site location. These state efforts have acquired and are using military installation geospatial information in their systems.

Military Health

I&E geospatial data assets have also been used to help plan, manage, track, and assess military health assets and potential health threats. This includes analyses at installations to national clinic and MTF capacity analysis, planning and management; to disease vector analysis and prevention; and medical emergency planning and response. For example, I&E geospatial data assets have been used to help develop emergency response plans and training for medical emergencies, such as a bioterrorism attack using anthrax or smallpox. Installation GIS data have even supported soldier physical training, as is discussed in the MWR section of this appendix.

I&E geospatial data assets have been especially useful in two main areas: disease and disease vector analysis and prevention, and providing medical care—medical capabilities planning and assessment. We discuss each of these areas in more detail.

Disease and Disease Vector Analysis and Prevention

I&E geospatial data assets are used to help detect, track, assess, plan, manage, and prevent the spread and potential spread of infectious diseases and the disease vectors, such as rodents and mosquitoes, at military installations, for a region, and throughout the United States and world. The spread of infectious diseases, especially by soldiers who are deployed and return from all over the world, has an important geographic component that makes geospatial analysis critical to helping to detect, treat, and prevent such occurrences. More and more geospatially based infectious disease surveillance and assessment systems are being developed to help detect and prevent the further spread of infectious diseases, whether from natural causes or from biological warfare. Such infectious diseases can range from the common cold and flu, to West Nile virus, malaria, and severe acute respiratory syndrome (SARS), to sexually transmitted diseases, and to the effects from potential biological terrorist attacks. The systems could help warn the medical community about the spread of the bird flu and its potential to become a pandemic.

Geospatial-based disease surveillance and assessment systems are being developed, tested, and implemented across the world. We prevent five examples here: three installation examples in the United States and two examples for the world. It is important to note that such geospatial applications have relevance for the warfighting mission as well and some are already used to help support warfighting medical concerns, such as detection for biological weapons use.

At two USAF installations, USAF medical staff members have used I&E geospatial data assets to examine and help prevent cold and flu epidemics on the installations. They assessed the locations of office buildings where many workers seemed to be catching colds and flu viruses. Then they targeted those areas for disease prevention, such as by providing education and training to the workers about frequent hand-washing and other preventive measures.

A GIS-based infectious disease surveillance system was developed and tested at Fort Bragg, North Carolina. The system was used to track the locations of two sexually transmitted diseases—gonorrhea and chlamydia. The system was found to be useful in developing preventive interventions.[12]

At APG, installation geospatial data have been used to help track and assess the locations of Lyme disease and West Nile virus. GIS staff and USACHPPM have conducted GIS based tick studies to examine the infection rate and location of Lyme disease. Field-sampling data about ticks' locations and disease rates were entered into the GIS by GIS staff. USACHPPM used these data to identify patterns and Lyme disease prevention measures. Similarly, sampling studies of mosquitoes with West Nile virus have been located digitally using GPS and analyzed with the GIS at APG.

The USAF Surgeon General Modernization Directorate has developed a GIS-based computer application, called Community Health and Medical Program, that provides integrated geospatial disease sur-

---

[12] For more information, see Zenilman et al. (2002).

veillance and outbreak detection using the Composite Occupational Health and Operational Risk Tracking data. This modeling system uses information about military patient locations (including military treatment facility locations) and influenza-like illnesses to help identify initial biological warfare symptoms that may initially look like the flu. It can also be used to help track influenza outbreaks, such as the potential spread of avian flu. Any illness clusters and trends that look suspicious are highlighted on the GIS map in red for military experts to drill down into and further examine.

Other parts of DoD have also been developing infectious disease tracking tools. The Office of the Deputy Assistant Secretary of Defense for Force Health Protection & Readiness (FHP&R) has a tool for early detection of infectious diseases at military treatment facilities that uses GIS technology and GIS data about military medical facilities. This tool is part of the Department of Defense-Global Emerging Infections System (DoD-GEIS) and is called the Electronic Surveillance System for the Early Notification of Community-based Epidemics (ESSENCE). Surveillance of syndromes recorded at the time of patient visit instead of specific diagnoses reported after laboratory or other diagnostic procedures can greatly lessen the time it takes to determine that an outbreak is occurring. In April 2002, the office created a new version of ESSENCE, which includes a web-based display of information that enables the user to drill down and obtain more information for each surveillance site.[13]

Providing Medical Care: Medical Capabilities Planning and Assessment

DoD provides both peacetime and wartime medical care to over 9.1 million beneficiaries, including soldiers (active, eligible Reserve, and eligible Guard), their families, and military retirees. I&E geospatial data assets are used to help make sure that the right medical capabilities are in the right place to meet demand.

---

[13] For more information on the ESSENCE tool, see *Electronic Surveillance Sysem for the Early Notification of Comunity-Based Epidemics* (n.d.).

OSD Health Affairs TMA/Health Programs Analysis and Evaluation Directorate has been developing a "Military Health System Atlas"—an atlas of military medical capabilities and their populations. It includes the military treatment facilities and different military populations (active duty, military families, retirees, etc.). It also includes information about military beneficiaries (both those within the United States and those abroad) as well as American Hospital Association and American Medical Association data, such as the locations of civilian hospitals. This system is used for multiple purposes:

- To help with resource allocation decisions, for example, to help determine if the right military medical resources are in an area or if there is a need to contract out with civilian facilities
- To help assess the health status and characteristics of different populations; for example, are there more smokers or obese people in certain areas
- To look at mission movements, such as BRAC realignments; for example, to determine how to support new missions and whether military or civilian MTFs have enough capacity.

Another example that is not yet being implemented is a system to support a Service Surgeon General's office, such as the USAF Surgeon General's office. The Surgeon General's office is interested in using geospatial data assets to help support business decisionmaking, such as using geospatial information to look at how well the MAJCOMs are meeting their medical facility business plans. A key performance metric is patient throughput at each medical facility. If this type of information was linked into a GIS, it could be useful for helping assess which medical facilities need to be focused on. For example, if the total number of monthly primary care visits for each facility was in the GIS along with the business plan expectations, then the system could compare the two. Each medical facility would be shaded red if the monthly visits were a large percentage below the values of the business plan, yellow if visits were a smaller percentage below the business plan, and green if at or above business plan expectations. Then the staff in the Surgeon General's office could focus on the reds to find out why there is such a disconnect between the actual and expected workloads. Similarly, information about the trends in enrollees at each MTF could be displayed in the GIS to help assess expected workload trends.

Morale, Welfare, and Recreation: Enhancing Quality of Life

I&E geospatial data assets also are used to enhance the quality of life of the U.S. defense community, which includes soldiers (active, Reserve, and Guard), their families, civilian employees, military retirees, and defense civilian staff. This mission area refers to any morale, welfare, and recreation (MWR), and quality-of-life activities from housing and medical services to family, child, and youth programs to recreation, sports, entertainment, travel, and leisure activities. I&E geospatial data assets are used to more effectively and efficiently develop, build, manage, and operate installation facilities and grounds and the services that are provided to the U.S. military community and to improve morale-building events and activities. They help improve the quality of the facility and grounds design and maintenance; the quality of service delivered; and the quality of MWR events. The assets are also used to help assess the quality of life across installations, such as the Army Sustainable Installation Regional Resource Assessment (SIRRA) tool. The USACE developed this GIS-based tool for BRAC to conduct a national-level comparison of Army bases by looking at quality of life, environmental, social, and economic parameters around bases.

Improving Installation Facilities and Grounds

Many installations use I&E geospatial data assets to improve the installation's housing, office buildings, recreation facilities, and grounds. A range of facilities—from golf courses, swimming pools, ball fields, military family housing, dining facilities, and jogging trails—are being designed, implemented, maintained, and operated with the help of I&E geospatial data assets. For example, at Fort Benning, GIS is being used to help assess where to place Residential Community Initiative (RCI) housing projects. RCI is an Army-wide public-private partnership program for providing high-quality cost-efficient housing on the base. GIS was used to evaluate possible areas by examining quality-of-life factors and desirable building conditions including topography, wetlands, and RCW habitat data to assess environmental impact, relationships to schools, noise from nearby training ranges, and relationships to other installation amenities, such as dining facilities and the BX.

I&E geospatial data assets have also been used to help plan for installation staff's dining needs. At NAS Patuxent River, the installation GIS was used to assess staff dining facility needs and determine the best location for a new cafeteria. In 1998, when several thousand new personnel were being added to the base because of the 1995 BRAC, the commander asked, where will all these people eat? Since many of the new employees would be commuters from the District of Columbia, lunch facilities, i.e., cafeterias, needed to be available within walking distance of their employment locations. Base planners were adding 2.0 to 2.5 million square feet of office space for the new staff and had to decide where to place a new cafeteria. To determine this, the GIS shop analyzed the existing and potential eating locations and created walking buffer zones showing the location of the new staff in relationship to the existing and potential eating locations.

The USAF uses a GIS-based application to improve both the aesthetic quality and maintenance of trees at Air Force installation. A GIS application called the Urban Tree Information System (UTIS) is used for urban landscape management. Tree locations have been entered into UTIS with the help of GPS and then integrated with other installation GIS data to identify ground maintenance problems and insect damage or disease that need to be addressed. Resource managers use UTIS to more easily identify high-priority tree planting sites and site-suitable species, because they can quickly visualize specific characteristics of the area. When used in conjunction with aerial photography, UTIS also provides "visual work orders" that clearly show tree maintenance and removal priorities. The UTIS database becomes the statement of work for grounds maintenance contracts and can also be used by the contractor and Quality Assurance Evaluator to document contract performance.

Improving Services That Are Provided to the Military Community

I&E geospatial data assets also are used to help improve the services that are provided to the U.S. military community. Such services include child development centers and schools, medical care, social services, libraries, shopping, and other services that military personnel and their families expect at installations and use to enhance their quality of life. Such services can be at an installation, region, functional command, headquarters, or OSD (an example is OSD using I&E geospatial data assets to better match medical resources to medical demand, as discussed in the military health section).

At Aberdeen Proving Ground, the Garrison GIS staff even helped the commander investigate an on-base home day care provider. A small child had wandered away from the day care site and had been found some distance away. The day care providers claimed that they had turned their backs only for a minute and the child was gone. The commander asked the GIS staff to analyze the situation. The GIS analysis showed that the child had been gone at least an hour and had climbed a fence and traveled a long way. The commander used this information to help make a decision about the day care center's license.

Even military religious services can be improved with the help of I&E geospatial data assets. The U.S. Army IMA Europe Chaplin Office wants to use I&E geospatial data assets to better provide religious services to Army soldiers stationed in Europe. Because of the restructuring of Europe IMA and realigning garrisons in Europe, the garrison chaplains need to decide if and where to build new chapels and where to place chaplains of different religious denominations. They would like to see a spatial distribution of soldiers by religious denominations and by new garrison locations to help them assess the best way to meet this demand with the appropriate chapels and chaplains.

A regional cross-service example of I&E geospatial data asset use occurs in Europe where such assets help provide educational services to U.S. military families. DoD uses GIS for facilities management of over 120 DoD schools in Europe. These schools are designed as campuses with several buildings. The GIS is used as a space management tool. DoD also runs "what-if" scenarios to examine what they would need to do if student populations fluctuated dramatically. For example, if there is a large increase in the number of students, the system can help identify the types of classrooms they need, ways to rearrange the space they already have, how many new teachers they need to hire, and can prepare detailed site maps and floor plans. School planners can run GIS queries to locate open spaces on campus by times of day.

Morale-Building Activities and Special Events

Morale-building events and activities are also designed, developed, and implemented with the help of I&E geospatial data assets. Such events can include special physical training activities, such as marathons; special public events, such as air shows; and other types of MWR activities. Providing maps to the participants is a usual part of such support. We present four examples here.

At APG, GIS data have supported soldier physical training. GIS staff members help assess where to place swimming, bicycle, and running venues as part of the triathlon event in the "best warrior event" of soldiers training. They also supplied maps for this event.

At Fort Hood, Texas, the MWR office wanted to develop a horse-riding route for a MWR program with a local horse club. GIS was used to develop the horse trail route around a recreation area with a scenic lake and provided the maps for the program.

At Cannon AFB, New Mexico, installation geospatial data are used to develop and support air shows. Key locations, such as emergency operations center, air show routes, and parking locations, are plotted in the GIS and used by diverse installation personnel including air show planners and emergency response personnel.

Often the installation support for such activities involves supporting more than just the MWR office as is illustrated with Camp Pendleton. At Marine Corps Base Camp Pendleton, California, geospatial staff have used I&E geospatial data assets to help support the Marine Corps Community Services (MCCS) staff in event planning. For example, staff members produce maps and do analysis for the 4th of July concert on the beach, the Ironman competition, and the mud run (a marathon through the mud that is open to the public). They help assess and show where to place parking and traffic routes and do "what-if" scenarios for different types of emergency situations. This information is used not only by the MCCS but also by military police, emergency responders, and other relevant installation staff.

Production of Installation Maps

Installation and environmental geospatial data assets are used to produce installation maps for a variety of purposes: providing installation navigation and directions for use on and near installations, for use in reports and official documents, to support training, and for other operational support. We briefly discuss each of these areas below. In addition, as discussed above, geospatial assets are used to produce official installation maps that are needed for legal and official reporting processes, such as maps showing a new training range area in a NEPA document for Fort Benning, Georgia or emergency response routes in an Emergency Operation Plan for NAS Patuxent River, Maryland.

Maps for Navigation and Directions

Geospatial data assets are used to produce maps for installation navigation and for people who need directions to navigate around or near an installation, whether in a car, a ship, or a plane. Such navigation support can be for military operations, such as using installation GIS data to create maps with navigation points for different installations. For instance at NAS Jacksonville, Florida, installation geospatial data assets are used to provide critical navigation points on an airfield and to provide airspace mapping for the air traffic control staff. They can also support installation staff, contractors, and other visitors who need to find their way around an installation. At Marine Corps Base Camp Pendleton, California, staff members produce a "Base Atlas," a hard copy Thomas-Brothers–like map book of about 100 pages to help direct contractors and others around the installation.

Requests for directional maps are so common that many organizations have developed specialized map products or geospatial web services to meet such needs. An installation example occurs at MCAS Cherry Point, North Carolina. Staff members at Cherry Point use their web-based GIS system to help visitors find their way around the installation. Installation GIS staff are planning to train visitor's center staff on how to use the system so that they can easily identify a visitor's destination and print out maps highlighting the route. A regional and functional example occurs with USAUR ITAM program, which publishes a catalogue that includes a set of 44 standard directional maps for key Army military installations throughout Europe.

An important need for maps arises when people from different organizations must work together in a battlefield, emergency response, or homeland security mission. In such operations, maps are needed that are consistent. Using the same grid reference system is important for coordinating and communicating location information. In combat areas, all U.S. military Services use the military grid reference system (MGRS), which is consistent with the U.S. national grid system (NGS) in the United States. This grid system is accessible by current commercial GPS systems as well as by military GPS systems, which make it easy to find locations in the field when the user has a map with the NGS on it. The MGRS is also on every official NGA-certified Military Installation Map. This single grid system can be very useful for coordination during homeland security if all organizations are using this national standard. In 2005, the USMC, U.S. Army, U.S. Northern Command (USNORTHCOM), and USGS have agreed to use this common grid reference system for homeland security. At the installation level, Marine Corps Base Quantico, Virginia, has used the MGRS/NGS grid system on its installation map. The USMC plans to use this grid system at all installations because of the benefits of using a standard system. The Army already uses this system.

The MGRS system is also being used by allies. The Germans use it on topographic maps. Because of the many benefits of this common reference system, Spangdahlem Air Base, Germany, uses MGRS for its crash grid.

Maps for Training

Preparing official maps for installation training had historically been an NGA mission. A certified Military Installation Map (MIM) is an installation map developed to NGA standards to support U.S. warfighters in training at installations around the world. Through 2005,

NGA acquired individual geospatial installation data from installations to create these special products, which take six months to one year to create. After NGA develops the official map, Defense Logistics Agency (DLA) stamps it, and then NGA prints hundreds or thousands. By the summer of 2005, NGA had a backlog of 107 U.S. requests for MIMs.

NGA does not want to produce them any longer. DISDI is working to have the installations produce these maps with NGA still certifying them and mass-producing the hard copies. NGA is working with Camp LeJeune to create the maps locally. If this experiment works, DISDI and NGA will try to expand it.

Since NGA was not producing training maps to meet Army demand, the U.S. Army ITAM program developed a substitute, the "Fort X Special" Map Product, so that it could produce its own maps for installation training purposes. It is very similar to an NGA MIM with the same scale and symbols. The Army also developed the Military Installation Map Template (MIMT) as a tool to quickly build these military installation maps, including mostly training range information.

Maps for Other Operational Support

At the installation level, geospatial data assets have been used to develop a wide range of standard map products to meet standard installation needs in all types of other installation operations from environmental and infrastructure support to security and emergency response. For example, at NAS Patuxent River, staff members use their web-based geospatial portal to provided standard maps or customized base maps for different applications for diverse business functions. In 2005, they provided 1,009 such map products. Maps were for the installation core business areas, including

- air operations support (38 maps)
- operations support (22 maps)
- personnel support (two maps)
- housing (four maps)
- facility support (143 maps)
- environmental (595 maps)

- public safety (175 maps)
- command and staff (30 maps).

I&E geospatial data assets have also been used to help provide map products for Service headquarters and OSD organizations. For example, the HAF GIO used IVT installation imagery data in the trip books for the Chief of Staff of the Air Force to use when visiting a base. This IVT imagery and other IVT data were also used to develop map products that supported OSD's BRAC process, which was discussed in detail in Chapter Five.

Public Affairs/Outreach

Public affairs/outreach includes the diverse ways that geospatial data assets have been used in public processes and for public outreach. Geospatial information in the form of maps, official documents, statistics, and web sites are used by public affairs, environmental, GIS, and other installation staff to inform military personnel and their families, surrounding communities, other parts of the United States and foreign governments, and the general public about installation-related issues. All sorts of public information—from basic installation travel conditions to formal environmental public outreach processes to congressional inquiries—is supplied with the help of geospatial data and analyses.

To illustrate these diverse outreach activities, we have organized the examples by three key audiences:

- military personnel, their families, and installation support staff
- surrounding communities and the general public
- Congress.

These audiences and outreach to them are not mutually exclusive; there is often overlap. However, for discussion purposes, these categories illustrate the diverse outreach missions that geospatial information support.

Military Personnel, Their Families, and Installation Support Staff

Geospatial data assets are used to provide military personnel, their families, and staff who work at military installations information about the installation, such as logistical and travel conditions on base, and to educate them about proper procedures and good citizen practices while on base, such as emergency response, security, and environmental concerns.

For instance, a common outreach use of geospatial assets is to notify people who work or live on a U.S. military installation about road or building closures or other activities on the installation that may interfere with people's activities. For instance, Ramstein AB, Germany, uses the base Emergency Response Tool Suite to show road closures, because the base experiences so many road closures from construction and other maintenance activities. It is also used for releasing road closure information to the public.

Another common application of I&E geospatial data assets is to provide cultural and environmental education to the U.S. troops and their families. For example, the USAREUR ITAM program developed the Sustainable Range Awareness (SRA) Viewer to educate soldiers and provide outreach about environmental issues to make them more aware of their behavior. It includes training videos that use installation GIS vector and imagery datasets. Such tools help to educate troops and their families as well as other audiences, such as at Camp Butler. As briefly mentioned above, Camp Butler environmental GIS staff created a 3-D Okinawa environmental educational video to educate troops and as a public relations demonstration for the camp. It is a 3-D island fly-over demonstration with narration and music about the history, culture, and environment of Okinawa. This attractive Hollywood-style video will educate viewers about the importance of the rich and diverse environmental and cultural resources on this subtropical island. It is primarily oriented toward U.S. troops and their families, but it also can be used to help with public outreach to others, such as visitors and those at public conferences in the United States.

Geospatial data assets are also used in outreach to military installation populations by educating them about the use of such resources. For instance, U.S. installations celebrate "GIS Day" with GIS Day open houses to help educate civilian and military communities about the application of GIS technologies, as Eielson Air Force Base, Alaska, has done.

Surrounding Communities and the General Public

I&E geospatial data assets are shared with local communities and the public to promote good relationships, help with communications about community concerns, and in official outreach processes.

Many installation activities require public and local community outreach by law and by policy, especially in the environmental and safety areas. For example, NEPA documents, official Restoration Advisory Board (RAB) meetings, and the development of an installation INRMP are all examples of where I&E geospatial information, usually in the form of maps in documents and on display, has gone to local communities and to the public at large. Noise outreach as discussed above is another environmental area where the use of I&E geospatial data assets have been useful for community outreach.

Because of safety concerns, geospatial information is also shared with local communities, such as runway APZs, which can extend off installations. For example, geospatial data showing the APZs for all the Naval Air Stations in Florida, including NAS Jacksonville, NAS Whiting Field, and NAS Pensacola, have been shared with nearby local governments. These local communities use this information to help plan appropriate land-use zoning near the APZs, such as by not allowing residential development in those areas.

U.S. military public affairs offices also use geospatial data assets because of political and public relations concerns with local communities and host nation governments. A good example is the April 2005 range fire in the CTA at Camp Butler, which was discussed above. Because of the threat of erosion on Okinawa, this fire was a concern to the local communities. Initially, the camp public affairs office released incorrect information about the size of the burn area; then, using the base GIS, the correct area was calculated and the information released to the public. Because two different numbers were given out, there was confusion about the incident and the base received negative press. Now Camp Butler has a new policy; the public affairs office will not release any such estimates until the burn area has been calculated in the GIS. The geospatial analyses of the incident and Camp Butler's control measures were also used to help explain the erosion effects and control efforts to the local public.

I&E geospatial data assets are also used and shared with the public when installations are being "good citizens" and participating in community activities, such as in the 2005 Boy Scout Jamboree, which was at Army Fort A.P. Hill, Virginia. The Boy Scouts used GPS receivers and installation maps to help teach them advanced navigation skills.

Congress

I&E geospatial data assets are also used to respond to official congressional requests and to help with communications regarding installation activities. Such applications occur at local installations, regional, and headquarters levels. We provide examples at each level.

An example of a local installation using geospatial data assets to answer a congressional request occurred at NAS Patuxent River. Congress inquired about the plan to place the presidential helicopter (VXX) at this installation. With 45 man-hours of work, NAS Patuxent River GIS staff conducted an analysis and created a series of 15 maps showing proposed sites and the constraints associated with them. Geospatial constraint information used in this analysis included airfield clear zones (ACZs), Military Construction (MILCON) project locations, no-build zones, wetlands, archaeological sites, IR sites, and ESQD arcs.

At a regional level, the Navy EFD South used installation GIS data to provide public informational maps for the congressional committee hearings on BRAC.

Service headquarters organizations use geospatial information when responding to congressional requests, ranging from providing simple maps to answering questions about military activities and their relationships to proposed legislation. Service headquarters organizations often produced state and national maps showing which installations are in that state or in each state in the country. ACSIM produced for the Army Office of Congressional Liaison 50 state maps that show point locations of Army installations with congressional district outlines. Similarly, USAF headquarters air and space operations branch staff produced a number of maps by state that show U.S. military installations and airspaces. A map for South Carolina shows DoD ranges, special-use airspaces, and USAF and Army bases. These maps were produced for a member of Congress or a state senator.

I&E geospatial data assets have even been used to help assess the feasibility of a congressional bill. Several years ago, a member of Congress wanted to create a bill that would prevent a federal agency from changing operations on any land it owned next to a national park, unless the agency first consulted with the NPS. The congress member asked the Services if they had any concerns about such a bill. USAF headquarters operations branch produced a GIS map showing the relationship between all 387 NPS properties and DoD properties. Many DoD properties, including the Pentagon, were next to NPS properties so the idea of such a bill was dropped.

Safety and Security

I&E geospatial data assets are also used to help installation and other military security forces and safety staff plan, assess, track, mitigate, and respond to safety and security concerns, which has helped to save lives and property. Specific mission functions supported include antiterrorism/force protection planning, explosive safety, electrical and other utility safety issues, operational safety, and public safety. Operational safety includes aircraft flight safety and weapons testing and training safety. Public safety includes policing, crime analysis, fire department support, disaster planning, and emergency medical services support. Some public safety examples were discussed above in the section on emergency planning and response, so they are not repeated here.

Examples are organized by the following four categories:

- antiterrorism/force protection
- explosives safety
- operational safety
- installation public safety and security.

Antiterrorism/Force Protection

I&E geospatial data assets support antiterrorism and force protection tracking, analysis, planning, and training exercises. Before presenting examples, it is important to understand official OSD definitions. According to the DoD Joint Publication 3-26, antiterrorism means "defensive measures used to reduce the vulnerability of individuals and property to terrorist acts, to include limited response and containment by local military forces."[14]

Force protection means:

> actions taken to prevent or mitigate hostile actions against Department of Defense personnel (to include family members), resources, facilities, and critical information. These actions conserve the force's fighting potential so it can be applied at the decisive time and place and incorporates the coordinated and synchronized offensive and defensive measures to enable the effective employment of the joint force while degrading opportunities for the enemy. Force protection does not include actions to defeat the enemy or protect against accidents, weather, or disease.[15]

At APG, GIS maps and analysis have been used extensively for force protection. After September 11, GIS staff conducted extensive analyses and produced over 1,000 maps in a week for force protection concerns. GIS analyses included the placement of guards and other security forces, the location for building hardened structures, helping assess how to improve security gates, and the mapping of buffer zones. Map examples include special maps around barracks and schools to help assess the protection of these facilities.

Installations and Service regional offices also develop special GIS-based tools to support antiterrorism/force protection activities. For example, the U.S. Navy Region Japan Public Work Center has developed and uses a GIS-based AT/FP planning tool. The AT/FP Tool allows security personnel to locate 12 AT/FP features anywhere on

---

[14] Department of Defense (2005b, p. GL-5).

[15] Department of Defense (2005b, p. GL-8).

the base map. Features include access point barriers, vehicle inspection areas, centralized parking, fixed and mobile posts, emergency staging areas, K9 locations, reaction force locations, command and control nodes, exterior personnel alerting systems, security cameras, and perimeter intrusion detectors. Positions for all features are set for each of four force protection conditions (Alpha, Bravo, Charlie, Delta). Security personnel can also assign AT/FP attributes to existing features, such as buildings and ships, identifying blast compliance values, setting standoff distances, etc. A risk component allows users to assign threat, vulnerability, and likelihood of attack values to any feature to assess the risks of attack.

Similarly, MCB Camp Pendleton public works office uses a GIS tool to help with AT/FP planning, analysis, and management. This GIS tool allows users to set/remove perimeters for buildings, select buildings by use type, set force protection levels base-wide, save settings, and apply threat multipliers. They also have barrier tools that allow the user to add/edit/delete barriers, select types and materials, and create reports.

Other parts of the department of defense also develop and use geospatial tools to help with the AT/FP mission. For example, CATS is a consequence management tool developed by DTRA that employs a suite of natural and technological hazard models to estimate and analyze effects from such phenomena as hurricanes; earthquakes; and chemical, biological, radiological, nuclear, and explosive events. CATS is used by military and civilian organizations to help assess, train, and plan for potential terrorist events as well as natural disasters. Installations, such as MCB Camp Lejeune and MCAS New River, load I&E geospatial data into CATS and use it to help plan and assess the consequences of natural and man-made disasters and for antiterrorism and force protection planning.

Explosives Safety

Since the U.S. military has to manage and store large amounts of explosive munitions, explosive safety programs are an important function at installations throughout the world. I&E geospatial data assets are used to support explosive safety planning and analysis activities by individual installations, the Services, and OSD. Specialized GIS tools, Explosive Safety Site Plans, and explosive safety surveys are examples discussed here.

ASHS is a GIS-based application software tool to assess capacities for explosive safety and for explosive hazard reduction. ASHS has been used for at least 80 USAF bases worldwide and at a few Army installations. For example, in 2002, ASHS was used at Osan Airbase, Republic of Korea, to identify and quantify threats and operational restrictions posed by the presence of munitions stocks and to recommend how to mitigate these threats and restrictions. ASHS has also been used to support the warfighting mission, such as when PACAF used it at host nation installations to support operations in Afghanistan. Mission effects can include:

- ability to store more munitions
- sometimes improving safety by reducing the hazards that personnel and assets are exposed to by our own munitions; however, the use of ASHS may potentially increase the overall safety risk just because more munitions may be stored at a site
- improved flow of munitions operations inside the storage area and flight line
- improved flow of sorties because of more efficient munitions flows
- more acreage made available for other activities on base by reducing the clear zones on-base and off.
- improved CIP data quality by validating the data, improving them, and returning them to the installation.

However, not all the benefits are accrued with every application. Application benefits and effects differ from location to location, because each installation has its own specific mission and local characteristics. Follow-on studies have shown less effect because local commanders have already initiated some or most of the necessary actions to mitigate or eliminate the risks from explosives to personnel and essential resources. For example, during Operation Noble Anvil, ASHS was used to provide warfighters with solutions to mitigate risks to resources after the initial force beddown and under the stress of surge operations.

As discussed in Chapter Three, every U.S. military installation, including permanent and contingency bases, is required to develop Explosive Safety Site Plans that describe and show how the installation meets DoD explosive safety standards. These plans must include maps, and I&E geospatial data assets are used to help develop them. For example, ASHS is used to help develop such plans. The USMC and U.S. Navy use a similar GIS-based software tool, the Explosive Safety Siting (ESS) Tool to help with Explosive Safety Site Plans. ESS has been used for this purpose at Kings Bay Naval Submarine Base and at NAS Fort Worth, Texas.

As discussed in Chapter Three, the DDESB reviews installation site safety plans to ensure that they comply with the DoD standards. DDESB staff members use recent imagery of an installation and installation boundaries to help review site plans. I&E geospatial data assets help them identify risks or violations in site plans. For example, the DoD explosive safety standard requires that explosives be kept a set distance from occupied buildings whether inside or outside the installation boundary. Because of encroachment around military installations, new off-base development can affect whether a site plan can be accepted. By viewing recent imagery, the DDESB staff members can view any new development near the installation and see if changes need to be made to the plan to meet the safety standard.

Operational Safety

I&E geospatial data assets have been used to improve the safety of military aircraft flight operations and ground training and equipment testing operations, helping to save both equipment and lives.

Many installations with flight runways, such as Langley Air Force Base, have a BASH and a Deer Aviation Safety Hazard (DASH) program because of the damage that such animals cause when they hit aircraft on or near installation runways. The USDA, Langley Air Force Base, and NASA Langley have been working together to monitor and manage the local osprey population. These birds, which were an endangered species and now are a species of concern in many states, have been nesting near runways and in runway approach areas. Birds are tracked through the Langley AFB BASH program and the resulting data are available through interactive maps generated and maintained by NASA's GIS team. The goal of the program is to minimize aircraft exposure to strikes while establishing an airspace environment where both aircraft and osprey can coexist relatively safely. GIS analysis of nest locations, bird strikes, and aircraft runway locations and approaches was used to help identify where to implement mitigation activities, such as changing the style of channel markers so that birds cannot nest on them and relocating fledglings. These activities have been successful because the osprey population has remained stable while the air strike hazards have been reduced; namely, fewer birds are nesting near the runways and aircraft approach areas.

NAS Patuxent River has used I&E geospatial data assets to conduct tree clearing analyses to improve runway operations safety. GIS staff there did a line-of-sight analysis for the radar tower to determine if trees were in the way. They mapped tree height for a new centerfield radar tower and calculated the minimum number of trees that needed to be taken down. In another case, pilots could not see runway lights because of tall trees, so GIS staff analyzed which trees needed to be cut for runway safety.

Similarly, at MCAS Cherry Point, North Carolina, GIS staff members used I&E geospatial data assets to help ascertain the operational safety of the ground training ranges. For example, they recalculated the small arms range fans using a submeter accurate GPS device. With the accurate range fans plotted, MCAS Cherry Point has started dialog with the required agencies to correct existing deficiencies pertaining to the Neuse River (a navigable waterway) and a small portion of the USDA FS Croatan National Forest. Using I&E geospatial data assets to ensure accuracy helps assess the safety effects of ground training ranges. This point will also be discussed further below in the training section to verify safety danger zones in range development and planning.

Geospatial data from outside an installation sometimes need to be combined with I&E geospatial data assets on the installation in operational safety assessments, such as analyzing an aircraft crash site outside an installation or near an installation boundary. For example, two military F-16s crashed off-base during an Air National Guard training exercise in Indiana. The Air National Guard needed GIS data outside the base to help provide situational awareness and analysis information for the crash site.

Installation Public Safety and Security

I&E geospatial data assets are routinely used to support installation public safety and security functions. Installation geospatial staff members work closely with fire departments, military police, and other security forces. Support is given to the base Emergency Operations Center and emergency 911 call and dispatching system, as discussed above, to base fire departments, and to military police. Providing maps for courts of law and as legal evidence is another security function supported by the assets.

Fire departments use I&E geospatial data assets to help with deployment analysis and dispatching. Such support was discussed above in the section on emergency response. However, we elaborate on two examples here. First, at NAS Jacksonville, Florida, and NAS Pensacola, Florida, installation GIS data are used to create fire hydrant maps showing the locations of all the installation fire hydrants. Second, at Fort Hood, Texas, the installation GIS system is even used to help to decide whether to put out a training fire. A training fire is extinguished only if it endangers personnel, equipment or habitat. Staff members plot active fires in the GIS to see where it is moving. GIS data are also used to help plan controlled burns.

Besides supporting dispatching systems, I&E geospatial data assets help support installation military police and other installation crime-fighting organizations by providing patrol car route, police stakeout, and crime investigation maps and by analyzing traffic accident patterns. We present an example of each. At NAS Patuxent River, GIS staff members create police patrol maps; and at APG, garrison GIS staff members assist the base police department by supplying them with police stakeout mapping support. In the latter case, the police have wanted to know what is in the line sight of a given radius for a given stakeout location. At Marine Corps Base Camp Pendleton, California, geospatial staff members have used I&E geospatial data assets to help support the Naval Criminal Investigative Service (NCIS). They provide NCIS with maps for use in such investigations as break-ins and for evidence in court. Camp Lejeune used I&E geospatial data assets to analyze the patterns of traffic accidents on base to help the military police develop speed limit changes.

I&E geospatial data assets also support installation security by providing legal evidence and maps about incidents on installations, such as legal boundary maps to be used in prosecuting trespassing fishermen and hunters who are caught on a military installation and claim that they were outside the base. For example, installation geospatial boundary information was used in court to show that protestors were on Vandenberg AFB property. One of many protesters who was arrested on installation property and tried in federal court tried to claim that he was not on Air Force property. An installation engineering assistant created a map using GIS data and used it in court to help support Vandenberg AFB's case. Similarly, at Marine Corps Base Camp Pendleton, geospatial staff members create and provide maps of traffic accidents to the military police to use as evidence in court. They even attach the military police photos to locations in the maps.

Strategic Basing

I&E geospatial data assets have also been used to support strategic basing decisions, including BRAC, and planning for major troop movements and installation changes throughout the world. For example, USAFE used installation geospatial data assets to help examine ideas about moving bases from Western to Eastern Europe. Similarly, Camp Butler used I&E geospatial data assets to help assess the effects of closing MCAS Futenma and relocating the base and troops to Guam because of encroachment around its runway and concerns by the Japanese government.

In the United States, the main strategic basing issue has been BRAC. In 2005, I&E geospatial data assets were used extensively to help in this process. In fact, a special tool, the IVT, was developed for it. IVT was the first major activity of the DISDI Office to help facilitate the sharing and use of I&E geospatial data assets and the first large-scale effort to share I&E geospatial assets from all Services in support of a high-level OSD process. For this reason, we conducted an in-depth analysis of this case study, as discussed in Chapter Five.

Training and Education of the U.S. Military

I&E geospatial data assets are used to support mission training and education, including joint, combined, and single Service training and education. Such assets have played a key role by providing field maps and web mapping systems for the commanders who plan and the soldiers who participate in training exercises; by helping to schedule and operate training ranges and classrooms; by helping to plan, assess, and build new air and ground test and training ranges; by helping to maintain training ranges; and by providing real world data for training simulators. At some installations, the use of such geospatial data assets is so common and mission-critical that one range officer at a large Army based stated that the base GIS data and system were used "like people use cell phones" and that it was difficult to imagine doing the job without them. The use of I&E geospatial data assets has improved training, saved money and time, helped to keep testing and training ranges open, and helped bases maintain operational flexibility.

To illustrate these diverse applications, examples are organized into three categories:

- installation training exercise and educational support
- planning, management, and development of training ranges and testing areas
- uses within the training system itself.

Installation Training Exercise and Educational Support

Installation GIS data have supported installation training exercises in several ways, including providing information for training orientation/land navigation, more active and timely maps, and interactive geospatial systems of base training areas, which allow soldiers to see an area ahead of time. For example, as discussed above, soldiers use the USAREUR ITAM ITAM Mapper and ITAM Viewer for training orientation and land navigation and to see an area ahead of time at Army training areas across Europe. Commanders also use geospatial applications to help plan training, as Army commanders do with the USAREUR ITAM ITAM Mapper and ITAM Viewer. Similarly, at Camp Butler, Okinawa, Japan, I&E geospatial data assets were used to conduct a terrain analysis for joint training exercises for the Jungle Warfighting Training Center.

Easy-to-use applications have been developed and placed in the field to provide timely support to range training exercises. At Camp Ripley, Minnesota, Army National Guard Training Site, a GIS-based kiosk was placed in the range control office so that soldiers can print their own maps of training areas (see the discussion and Figure 7.1 in Chapter Seven).

I&E geospatial data assets have also helped the USAF in pilot training. Instructors use the tactical pilotage chart as a supplemental tool when training F-16 student pilots at Luke Air Force Base near Phoenix, Arizona. The charts were developed using GIS and contain surface and airspace data, such as elevations, major roads and cities, airfields, obstructing towers and cables, training routes, restricted airspace, no-fly areas, communication points, tactical areas, and other military operating areas. More than 100 of these charts were distributed and used by fighter squadrons of the 56th Range Management Office.

I&E geospatial data assets are also used to help manage and schedule training and educational infrastructure. Navy EFD South has been working with the Naval Education and Training Command to use installation GIS data to track and manage training buildings, classrooms, and training equipment worldwide. So far, this system is used at NAS Pensacola, Florida, and at many other U.S. Navy installations as well in the Pacific. This example also shows how such assets are used by functional commands and across many installations to help support the training and education mission.

Planning, Management, Development, and Operation of Training Ranges and Testing Areas

Installation GIS data have helped in the planning, analysis, management, development, and operation of training and testing areas, especially for air, ground, and water ranges. Such analyses have helped keep ranges open and have helped maintain their operational flexibility. They also have helped increase training space and hours on a range. For example, at Fort Stewart, Georgia, the use of GIS to more accurately calculate surface danger zones has enabled the installation to use more of their land for training. We present diverse examples of such applications here.

I&E geospatial data assets have been critical in the planning and development of new training ranges, especially in addressing environmental and safety concerns. As examples, Fort Benning, Georgia, has added new training ranges and MCB Camp Lejeune, North Carolina, has expanded a rifle range. As mentioned briefly above, because of the 2005 BRAC, Fort Benning is adding seven new ranges and upgrading many others and GIS data are being used to help determine where to locate the new ranges. These assessments cover a range of effects, including environmental impact and the effect on existing training ranges. GIS is being used to calculate the SDZs—exclusion areas identified to protect personnel from weapons firing during training. Some of these training areas will need to be closed because they are in the SDZs. Noise contours are calculated to assess the noise effects on residential housing or communities off installation if the range is near the installation boundary. Geospatial data assets are also being used to assess the effects on wetlands, the endangered red-cockaded woodpecker, and other key species, such as the gopher tortoise. Geospatial data assets enable installation staff to perform this analysis more quickly and in a more automated fashion. Similarly, a new Special Operations Command is being located at Camp Lejeune. To accommodate this new command, a rifle range needs to be expanded, requiring a large amount of construction. The GIS staff members are assessing this expansion using the base GIS system to examine range development characteristics and key factors, such as the location of utilities, wetlands, cultural resources, and RCW habitat.

Sophisticated geospatial tools are also being developed to help support such training range analyses and share the capabilities more broadly across the training range community. For example, under the Army's Sustainable Range Program (SRP) an evolving GIS-based tool-set—the RMTK—has been developed for the range staff to assist in range planning and operations. It includes the Surface Danger Zone (SDZ), a Noise, an On-Range Munitions, and the Range Design and Planning (RDAP) Tools. We summarize three of these briefly below:

- The SDZ Tool is used to create surface danger zones for different weapon systems. It allows range safety officers and range control officers to interactively create SDZs by selecting weapon systems, target points, target media, and firing points. Users are able to identify firing and target locations in the map interface by entering MGRS coordinates or by selecting existing points. The tool also allows users to create SDZs for Combined Arms Live Fire Exercises (CALFEX) by combining individual SDZs.
- RDAP is a new tool used to place potential ranges and associated SDZs on an installation and then dynamically move them to see relationships with other data, such as environmental constraints and base boundaries. It has been beta-tested at Fort Bliss.
- The RMTK Noise Tool is a noise prediction and impact assessment software tool designed to enable range managers and trainers to quickly assess the noise effects of training or testing on any given day and for a variety of weather conditions. It also allows them to plan preliminary range siting for noise. Range Managers and trainers use this tool to improve the scheduling (time and location) of training or testing.

The RMTK has been used by a number of Army and USMC bases. The Army has even used it to help develop training ranges in Iraq, as was discussed in Chapter Four.

I&E geospatial data assets are also used to help plan and operate unique training and testing areas, such as in urban caves, which have become more important in recent years given our operations in Afghanistan and Iraq. For example, at Fort Leonard Wood, Missouri, caves are being surveyed, inventoried, monitored, and mapped to help assess their suitability for military training. Over 50 caves have been surveyed for biological and cultural resources over a two-year period. The inventory identified nine caves that are suitable for military training. The caves were recorded in a GIS database with the integration of photos of cultural and biological information. The project will also help develop "cave policy" for installations to facilitate management for cultural, natural, military training, and recreational resource issues.

This activity was funded by the OSD Legacy Resource Management Program,[16] established by congressional legislation in 1990 to provide financial assistance to DoD efforts to preserve our natural and cultural heritage. The program helps DoD protect and enhance resources while supporting military readiness and training. Many of the other Legacy projects[17] also use I&E geospatial data assets to help in monitoring and analyses.

A good example of how I&E geospatial data assets helped keep ranges open and helped maintain operational flexibility at those ranges occurred when the USAF conducted geospatial analysis to protect training airspace from commercial airline encroachment. In fall of 2005, after Hurricanes Rita and Katrina, Congress wanted to open up more of the Gulf of Mexico to oil exploration. One such area is "Lease Sale 181." Because of Eglin AFB training airspace, the USAF had some concerns about some of the areas. USAF headquarters operations branch in charge of airspaces and ranges staff produced a map showing current lease sales using Department of the Interior (DOI) mineral management division's (MMD) data on active leases, oil rig and pipeline locations, and the Eglin AFB airspace. The USAF worried about any commercial activity, such as oil rigs and windfarms, east of a north-south line marking the western edge of Eglin AFB's airspace. Oil rigs there could create safety concerns and concerns also about electromagnetic interference. Areas to the west of this line

---

[16] For more information, see Proffitt (2005).

[17] Such projects may involve regional ecosystem management initiatives, habitat preservation efforts, archaeological investigations, invasive species control, Native American consultations, or monitoring and predicting the migratory patterns of birds and animals.

are far enough away from airspace operations to cause no problems. The map prepared by operations branch was sent to the Air Force Chief of Staff, Florida, members of Congress, Eglin AFB staff, and other relevant individuals and was used to help protect the military's training airspace.

I&E geospatial data assets also support daily training operations, such as the Integrated Range Control System (ICRS) at MCB Camp Lejeune. The ICRS links the range scheduling database into the base GIS so that the training live fire desk has a geospatial view of ongoing training exercises. The system also includes radar so that real-time information about aircraft locations is also displayed. With this system, range operators can see live fire training locations, aircraft locations, SDZs, and other range information, giving them an integrated accurate view of ongoing activities. This system has helped improve the operational safety of training exercises.

Besides helping with operations at training ranges, I&E geospatial data assets are also used to help with both the daily and the long-term maintenance of ranges. One common maintenance example is research, monitoring, and analysis related to erosion at heavily and intensively used ground training sites. For example, Fort Huachuca's ITAM program has used a video mapping system to help track, assess, and manage erosion problems in training areas. Installation staff link a small pocket-sized GPS unit to a videocamera and film eroded training areas. The data are then entered into the GIS system using the GPS coordinates so that the user can click on selected roads or trails in a training area and see video of that location. The system is especially useful for inventorying erosion and gullying problems.

Uses Within the Training System Itself

Geospatial data are also used within training systems themselves, such as in installation flight and ground simulators, saving training range time and dollars. Often, such savings include cost avoidance savings from sharing geospatial data, especially high-resolution imagery. For example, at NAS Patuxent River, the sharing of installation aerial imagery in simulators saved an estimated $1.5 million in cost avoidance. The NAS Patuxent River "Man Flight Simulator" and the U.S.

Naval Test Pilot School used the GIS office 6-inch aerial imagery for the base. Each organization saved an estimated $0.75 million in cost avoidance by not having to purchase the imagery.

I&E geospatial data assets are used in many 3-D flight simulations for Army, USMC, and Navy installations, such as at Fort Benning, Georgia; Fort Irwin/NTC, California; Fort Bragg, North Carolina; Grafenwoehr, Germany; Combat Maneuver Training Center (CMTC) Hohenfels, Germany; Camp Pendleton, California; and Norfolk Naval Base, Virginia.

Such applications can even help reduce the amount of time that soldiers need to train on a range, which is an important mission effect, since training ranges face so many pressures, including encroachment, and the need for training range space is expected to increase in the future.[18] For example, at Fort Hood, Texas, the range GIS aerial and topographic data are used in tank and aviation simulators, which help orient the soldier and save valuable time on the training range. It has cut the amount of time that helicopter pilots need to spend on the gunnery range by about one-third. A-64 Apache helicopter pilots fly the Fort Hood model before using the gunnery range. Previously, they made an initial flight pass at the gunnery range and then two more passes to fly and actually shoot. However, with the realistic installation simulator, they no longer need to make that first pass, since any issues have been worked out in the simulator.

Even NGA-developed warfighting training tools and simulations use I&E geospatial data assets. For example, training staff members at Marine Corps Air Ground Combat Center at Twentynine Palms, California, use I&E geospatial data assets in an NGA GIS Warrior Extension–Terrain Tool, which facilities terrain analysis and visualization, such as line-of-sight visibility, and the construction of domes and fly-through simulations for installation training.

---

[18] Today, an Army Stryker brigade combat team has a doctrinal battlefield footprint of 40 x 40 kilometers (1,600 sq km) and the future force is expected, by one estimate, to have a 75 kilometer radius (17,671 sq km) doctrinal footprint requirement (Knott and Natoli, 2004, p. 12).

Transportation

I&E geospatial data assets are used to help plan, build, operate, and maintain transportation assets—whether air, road, rail, trail, or sea transportation. They are also used to assess and implement transportation routing and scheduling. We discuss these two categories below.

I&E geospatial data assets are also used to help support transportation safety as briefly discussed in the section on safety and security. However, an interesting example, not mentioned there, has to do with driving and road safety. At USAF's Aviano Air Base, Italy, I&E geospatial data have been used to reduce driving accidents on local winding roads involving airmen from the installation. The installation security forces used to write accident reports by hand, including only the accidents' general locations. Now, installation GeoBase staff members plot the accidents on a digital map and look for accident clusters to identify dangerous driving areas. This information is then given to the airmen to improve driving safety. The maps are to be included in a PowerPoint presentation to help educate new airmen about driving hazards.

Planning, Building, Operating, and Maintaining Transportation Infrastructure

I&E geospatial data assets are commonly used at installations to help plan, design, construct, operate, and maintain new and existing roads on the base. For example, at Fort Hood, Texas, GIS data and software are used to help with road design when rerouting a road or building a new one. Installation staff members use the data for planning the location of the road centerline, making sure the road does not exceed a 6 percent grade, and to place drainage ditches. In addition, geospatial data are shared with state, local, and other federal agencies to help with their road projects, such as at MCB Camp Pendleton, California. Camp Pendleton geospatial staff members share I&E geospatial data with the California Department of Transportation to use in planning, maintaining, and operating California Highway 5, which runs through Camp Pendleton.

In the base operations and management section above, we presented examples how I&E geospatial data assets also support runway and port planning and construction. Such assets also support the operation of air and water transportation operations. An example of how such data support sea transportation occurs with U.S. Navy Hawaii, where the installation web-based geospatial information system is used to publish and distribute detailed maps including water taxi schedules.

The USAF headquarters operations branch in charge of airspaces and ranges routinely uses installation geospatial information to provide other federal transportation agencies, such as the National Transportation Safety Board (NTSB) and the FAA, with airspace and flight operation information. For example, when the USAF returns control of airspace from a military training route to the FAA, branch staff members produce a map for the FAA.

The USAF, Air National Guard, and U.S. Navy all use geospatial data assets to help in the investigation of any fatal aircraft accidents both on and off base. For instance, NAS Patuxent River GIS staff members provide mapping support for such investigations on the base. A contractor uses GPS to locate the crashed aircraft parts and GIS staff members provide a site map showing the scatter pattern of the aircraft debris.

Route Planning and Assessment

Another common transportation use of I&E geospatial data assets is to plan, assess, and schedule transportation routes on, to, and from installations. Many installations use such assets to help plan emergency evacuation routes and ordnance and hazardous waste transportation routes. For example, at USAF's Aviano Air Base, Italy, I&E geospatial data have been combined with local road data and hospital location data to create a regional picture for the installation medical group. This system is used to help assess travel information during a medical incident when someone must be transported to the local Italian hospital.

As discussed in Chapter Four in the section on warfighting logisistics, SDDCTEA has developed IRRIS, a secure web-assessable GIS system to monitor transportation logistics data and real-time tracking information. This system is also used to help schedule and route military cargo throughout the world.

Bibliography

Aberdeen Proving Ground, "Aberdeen Proving Ground Geographic Information System Five Year Plan 2006–2010," draft, Aberdeen Proving Ground, Md., 2005.

Aldridge, E. C., Jr., Chairman, Infrastructure Steering Group, USD (AT&L), "Transformation Through Base Realignment and Closure (BRAC 2005) Policy Memorandum One—Policy, Responsibilities, and Procedures," memorandum for Secretaries of the Military Departments, Chairman of the Joint Chiefs of Staff, Undersecretaries of Defense, Director Defense Research and Engineering, Assistant Secretaries of Defense, General Counsel of the Department of Defense, Inspector General of the Department of Defense, Director Operational Test and Evaluation, Assistants to the Secretary of Defense, Director Administration and Management, Directors of the Defense Agencies, Washington, D.C., April 16, 2003.

Allred, Paul, "GIS Provides Transportation Logistics, Real-Time Tracking for U.S. Military," *U.S. Military Traffic Management Command Translog*, September 22, 2005, p. 26.

American Farmland Trust, *Agricultural Conservation Easements*, Fact Sheet, Washington, D.C., September 1998.

Application of Hyperspectral Techniques to Monitoring and Management of Invasive Weed, Strategic Environmental Research and Development Program, Conservation CS-1143, Arlington, Va., 2004.

Army AL&T, Army Logistics Sustainment: The Leap Ahead, Headquarters Department of the Army, September–October 2002.

Assessing the Impact of Maneuver Training on NPS Pollution and Water Quality, report brief, Strategic Environmental Research and Development Program, Compliance CP-1339, Arlington, Va., 2003.

Assistant Chief of Staff for Installation Management (ACSIM), *U.S. Army Base Realignment and Closure: Environmental Condition of Property Guide for BRAC 2005*, Army, July 8, 2005.

Baker, John C., Beth E. Lachman, Dave Frelinger, Kevin M. O'Connell, Alex Hou, Michael S. Tseng, David T. Orletsky, and Charles Yost, *Mapping the Risks:*

Assessing the Homeland Security Implications of Publicly Available Geospatial Information, Santa Monica, Calif.: RAND Corporation, MR-142-NGA, 2004. As of April 30, 2007:
http://www.rand.org/pubs/monographs/MG142/

Barnes, Scottie, "Southern Command's Haitian Routing," *GeoIntelligence*, May/June 2004, pp. 14–18.

Barthello, Marc, and Alan Beiagi, "Online Tracking from Port to Port," *Geospatial Solutions*, Vol. 13, No. 2, February 2003, pp. 28–31.

Bernhardsen, Tor, *Geographic Information Systems: An Introduction,* 2nd ed., New York: John Wiley & Sons, Inc., 1999.

Board on Earth Sciences, Licensing Geographic Data and Services, Washington, D.C.: The National Academies Press, 2004.

Camp Lejeune, *IGIR Data Catalog*, North Carolina, October 2005.

Carpenter, Bobby, "Web Based Tool to Simplify the Environmental Impact Analysis Process," *CADD/GIS Insights Bulletin*, Vol. 4, No. 1, Summer 2004, p. 6.

Center for Technology in Government, *A Framework for Evaluating Public Sector Geographic Information Systems*, CTG.GIS-005, Albany, N.Y., 1995.

Clark, Matt, "Expanding the Fort Rucker GIS Enterprise," *SRP Newsletter*, U.S. Army Sustainable Range Program, Issue 27, July 2005.

"Cleaning up Kaho'olawe: U.S. Navy Completes Massive UXO Project," *GeoIntelligence*, September/October 2004, pp. 10–11.

Cline, Melinda K., and C. Steve Guynes, "The Impact of Information Technology Investment on Enterprise Performance: A Case Study," *Information Systems Management*, Vol. 18, Issue 4, Fall 2001, pp. 70–77.

"Corps of Engineers Creates Maps for Hurricane Relief," *Geo World*, Vol. 11, No. 18, November 2005, p. 14.

Cullis, Brian J., and Hal Tinsley, "One Installation, One Map," *GeoIntelligence*, September/October, 2004, pp. 16–21.

Dalby, Learon, "GIS and the Clinton Library," *Geospatial Solutions,* June 2005, pp. 19–25.

Davenhall, Bill, "Building a Community Health Surveillance System," *ArcUser*, January–March 2002, pp. 18–21.

Davern, M. J., and J. Robert, "Discovering Potential and Realizing Value from Information Technology Investments," *Journal of Management Information Systems*, Vol. 16, No. 4, Spring 2000, pp. 127–136.

Defense Installation Spatial Data Infrastructure, Deputy Under Secretary of Defense for Installations and Environment, Business Transformation Directorate, *DoD Real Property Inventory, RPI Site and Land Recommendations for the*

Spatial Data Standards for Facilities, Infrastructure and Environment, SDSFIE, Installations and Environment Business Transformation Directorate, May 27, 2005.

Department of the Army Assistant Chief of Staff for Installation Management, *Data Standards for Computer Aided Drafting and Design, CADD, Geographic Information Systems, GIS and Related Technologies,* memorandum to the Department of the Army, Washington, D.C., October 16, 2001.

————, *Training Requirements for Geographic Information System, GIS Support,* memorandum to the Department of the Army, Washington, D.C., August 6, 2003.

————, *U.S. Army Geographic Information Systems Repository, GISR,* memorandum to the Department of the Army, Washington, D.C., March 31, 2005.

Department of the Army and the Air Force National Guard Bureau, *Cartographic Standard Requirements for GIS Products,* memorandum for State Environmental Program Managers and State Geographic Information Systems, GIS Program Managers, n.d.

Department of Defense, *Department of Defense Directive: National Imagery and Mapping Agency,* Number 5105.60, October 11, 1996.

————, *Data Sharing in a Net-Centric Department of Defense,* Directive 8320.2, December 2, 2004.

————, *Base Structure Report: Fiscal Year 2005 Baseline,* Arlington, Va., 2005a.

————, *Homeland Security,* Joint Publication 3-26, August 2, 2005b.

Development of an Adaptive Framework for Management of Military Operations in Arid and Semi-Arid Regions to Minimize Watershed and Instream Impacts from Non-Point Source Pollution, Strategic Environmental Research and Development Program, Compliance CP-1340, Arlington, Va., 2004.

Dickinson, H. J., and H. W. Calkins, "The Economic Evaluation of Implementing a GIS," *International Journal of Geographical Information Systems,* Vol. 2, No. 4, 1988, pp. 307–327.

Directorate of Public Works, *Geographic Information System Implementation Plan,* Aberdeen Proving Ground, Md., 1992.

Eglin Air Force Base Integrated Natural Resources Management Plan, 2002–2006. Natural Resources Management Branch, Eglin Air Force Base, Fl., 2002.

Electronic Surveillance Sysem for the Early Notification of Community-Based Epidemics, n.d. As of July 24, 2007:
http://www.geis.fhp.osd.mil/GEIS/SurveillanceActivities/ESSENCE/ESSENCEinstructions.asp

El-Swaify, Ayman S. A., "Navy PWC Yokosuka's Recipe for Building a GIS," *CADD/GIS Bulletin*, Vol. 00, No. 2, September 2000, pp. 1–4.

———, "Using GIS to Facilitate Anti-Terrorism/Force Protections, ATFP Planning," *CADD/GIS Bulletin,* Vol. 03, No. 2, Fall 2003, pp. 7–13.

Executive Order 13112, "Invasive Species," February 3, 1999.

Farley, Scott M., and Scott C. Belfit, "Addressing Encroachment with Cooperative Agreements and Conservation," *Federal Facilities Environmental Journal*, Vol. 12, No. 2, Summer 2001.

Federal Geographic Data Committee, "Guidelines for Providing Appropriate Access to Geospatial Data in Response to Security Concerns," Reston, Va: U.S. Geological Survey, June 2005. As of April 30, 2007: http://www.fgdc.gov/policyandplanning/fgdc-guidelines

Fort Future: Simulation and Modeling for Installation Transformation: Program Overview, U.S. Army Corps of Engineers, Engineer Research and Development Center, ERDC/CERL TN-037, September 2003.

Frank, George R. "The Fort Bragg Firebreak Sustainability Study," *SRP Newsletter*, U.S. Army Sustainable Range Program, Issue 27, July 2005.

Fuentes, Gidget, "Massive Model; Camp Pendleton 'Terrain' Built from Digital Data to Create Lifelike Quality," *Marine Corps Times*, July 28, 2003, p. 16.

"GeoFidelis: GeoSpatial Operations and Procedures," U.S. Marine Corps Review Draft, March 2005.

Gillespie, Stephen, *Cost-Benefit Analysis of the Initiative to Accelerate Digitizing of 7.5-Minute Maps*, Reston, Va.: U.S. Geological Survey, 1993.

———, "GIS Technology Benefits: Efficiency and Effectiveness Gains," Reston, Va.: U.S. Geological Survey, 1994a.

———, "Measuring the Benefits of GIS Use: Two Transportation Case Studies", *URISA Journal*, Vol. 6, No. 2, Fall 1994b.

———, "A Model Approach to Estimating GIS Benefits," Reston, Va.: U.S. Geological Survey, unpublished, 1997.

———, "An Empirical Approach to Estimating GIS Benefits," *URISA Journal*, Vol. 12, No. 1, 2000.

Gillies, J. A., W. P. Arnott, V. Etyemezian, H. Kuhns, H. Moosmuller, D. DuBois, and M. Abu-Allaban, *Characterizing and Quantifying Local and Regional Particulate Matter Emission from Department of Defense Installations*, Strategic Environmental Research and Development Program, Arlington, Va., March 2005.

Gould, Nani, "Army Engineers Help Restore Iraqi Wetlands," *Engineer,* October–December 2004.

Gould, Nani, and Fauwax Hanbali, "Army Engineers Develop Model to Support Iraqi Water Management," *Engineer,* October–December 2004.

Greenfield, Victoria A., Valerie L. Williams, and Elisa Eiseman, *Using Logic Models for Strategic Planning and Evaluation,* Santa Monica, Calif.: RAND Corporation, TR-370-NCIPC, 2005. As of April 30, 2007: http://www.rand.org/pubs/technical_reports/TR370/

Gutierrez, Samuel, "Launching Spacecraft from a Wildlife Refuge," *ArcNews,* Vol. 26, No. 1, Spring 2004, pp. 8–9.

Hall, J. A., P. Comer, A. Gondor, R. Marshall, and S. Weistein, *Conservation Elements of a Biodiversity Management Framework for the Barry M. Goldwater Range, Arizona,* Tucson, Ariz.: The Nature Conservancy of Arizona, October 2001.

Headquarters Air Force Geo Integration Office, *ACC GIO Data Delivery Receipt,* Arlington, Va., July 22, 2005a.

———, *U.S. Air Force, GeoBase Data Inventory Results 2005,* Arlington, Va., 2005b.

———, *Installation Mapping and Visualization Common Installation Picture 2006 Control Document,* Arlington, Va., March 2006.

Headquarters Air Force Geo Integration Office, Office of Civil Engineering, *USAF GeoBase Enterprise Architecture: Version 1.0,* Washington, D.C., January 2003.

———, *USAF Garrison Mapping Concept of Operations: Version 2.0,* Washington, D.C., June 2003.

Headquarters U.S. Air Force/ILE, *"DTRA Access to USAF CIP Data for JSIVA Program,"* memorandum to ALMAJCOM/CE, Washington, D.C., January 14, 2005.

———, *Sustainable Installations Regional Resource Assessment SIRRA Data Request,* August 31, 2005.

Helmstetter, Andrea, and Debby Atencio, "Eglin Air Force Base and Sea Turtle Nesting: A Success Story," University of Michigan, n.d. As of April 30, 2007: http://www.umich.edu/~esupdate/library/97.09-10/helmstetter.html

Houston, Robert, Sean Sinclair, and Jim Robertson, "Where's the Trash?" *ArcUser,* July–September 2005, pp. 66–69.

Huckkerby, Cheryl L., "Fort Hood, Texas: CRM in the Home of the Army's Largest Fighting Machines," *Cultural Resources Management,* Vol. 2001, No. 3, March 2001, pp. 13–16.

Hughes, D., "The Homeland Defense Picture: The Idea Is to Combine Military Tracks with Civil Data on Assets Such as Nuclear Plants and Transportation Grids," *Aviation Week & Space Technology,* Vol. 159, No. 12, 2003, p. 56.

Impacts of Military Training and Land Management on Threatened and Endangered Species in the Southeastern Fall Line/Sandhills Community, Strategic Environmental Research and Development Program, Conservation CS-1302, Arlington, Va., 2002.

Integrated Natural Resources Management Plan: Fort Benning Army Installation 2001–2005, Fort Benning, Ga.: Environmental Management Division, 2001.

Integrated Natural Resources Management Plan: Robins Air Force Base, Air Force, Robbins Air Force Base, Ga., September 2001.

ITAM GIS Regional Support Centers, *Geodatabase Concepts: Setting Topology Rules for ITAM GIS Datasets,* Army, April 1, 2004.

IVT Program Office, ODUSD(I&E), *Installation Visualization Tool, IVT: Charter,* Arlington, Va., September 2003.

IVT Program Office, ODUSD(I&E), *Installation Visualization Tool, IVT Application Users Guide,* Arlington, Va., June 2004.

IVT Program Office, ODUSD(I&E(BT)), *Installation Visualization Tool Operational Instruction: Version 1.0,* Arlington Va., June 1, 2004.

IVT Program Office Headquarters and Air Force Geo Integration Office, Office of the Civil Engineer DCS/Installations & Logistics, "DOD Installation Visualization Tool Quality Assurance Plan," Version 1.1, Washington, D.C.: Pentagon, December 31, 2003.

Jang, Mihwa, "Collaborative Effort Creates Online GIS; Data Partnerships," *Geo World,* Vol. 1, No. 17, January 2004, p. 32.

Johnson, Bill, *The NYS GIS Data Sharing Cooperative: An Innovative New Model for Data Sharing and Partnerships,* NYS Department of Transportation, October 2, 1997. As of April 30, 2007:
http://www.nysgis.state.ny.us/coordinationprogram/reports/datasharing/index/cfm

Keys, David, Heather Melchior, and Stacey Guthiel, "Implementing GIS Across an Army Major Command," 2002 ESRI International User Conference, July 8–12, 2002, San Diego, Calif., July 10, 2002. As of December 27, 2005:
http://gis.esri.com/library/userconf/proc02/pap0779/p0779.htm

Keysar, Elizabeth, and Hibba Wahbeh, "The Land-Use Deconfliction Process at Fort Lewis: Enhancing Integration," *Federal Facilities Environmental Journal,* Autumn 2000, pp. 105–115.

Knott, Joseph L., and Nancy Natoli, "Compatible Use Buffers: A New Weapon to Battle Encroachment," *Engineer,* October–December 2004.

Kroeker, Steve, "High-Tech Mapping Helps the Environment," Colorado State University, n.d. As of April 30, 2007:
http://gis.esri.com/library/userconf/proc99/proceed/papers/pap218/p218.htm

————, "Monitoring Erosion at Thunder Mountain," *GPS World*, Vol. 10, No. 5, May 1999, p. 348.

Kucera, Henry, "Hunting Insurgents: Geographic Profiling Adds a New Weapon; Crime Mapping," *Geo World*, Vol. 10, No. 18, October 2005, p. 30.

Kucera, Paul A., Witold F. Krajewski, and Yong C. Byron, "Radar Beam Occulation Using GIS and DEM Technology: An Example Study of Guam," *American Meteorological Society*, July 2004, pp. 995–1006.

Kuester, Greg, Al DeAngelo, and Gary Sheets, "Proving GIS Works at the Aberdeen Proving Ground," *Geo World*, Vol. 13, No. 11, November 2000, pp. 42–45.

Ledbetter, Mark, "Blueprints for a Citywide GIS: Scottsdale's Award-Winning System Provides a Profitable Example," *GIS World*, Vol. 10, 1997, pp. 62–68.

Lozar, Robert C., C. R. Ehlschlaeger, and Jennifer Cox, *A Geographic Information Systems, GIS and Imagery Approach to Historical Urban Growth Trends Around Military Installations*, U.S. Army Corps of Engineers, Champaign, Il.: ERDC/CERL, May 2003.

Lozar, Robert C., William D. Meyer, Joel D. Schlagel, Robert H. Melton, Bruce A. MacAllister, Joseph S. Rank, Daniel P. MacDonald, Paul T. Cedfeldt, Pat M. Kirby, and William D. Goran, *Characterizing Land Use Change Trends Around the Perimeter of Military Installations*, ERDC TR-05-4, Champaign, Il., April 2005.

Marshall, R. M., S. Anderson, M. Batcher, P. Comer, S. Cornelius, R. Cox, A. Gondor, D. Gori, J. Humke, R. Paredes Aguilar, I. E. Parra, and S. Schwartz, *An Ecological Analysis of Conservation Priorities in the Sonoran Desert Ecoregion*, Tucson, Ariz.: The Nature Conservancy of Arizona Chapter, Sonoran Institute, and Instituto del Medio Ambiente y el Desarrollo Sustentable del Estado de Sonora with support from Department of Defense Legacy Program, Agency and Institutional partners, April 2000.

Marshall, Robert, *Schedule of Planning Activities Through September 30, 1998 for the Sonoran Desert Ecoregional Plan*, Tucson, Ariz.: The Nature Conservancy, 1998.

McInnis, Logan, and Stuart Blundell, "Analysis of Geographic Information Systems (GIS) Implementations in State and County Governments of Montana," prepared for the Montana Geographic Information Council, Helena, Mont., October 1998.

McSherry, Patricia, and Michael Hardy, "Hurricane Preparedness," *Military Geospatial Technology*, January 2006.

Mitchell, Lee E., "Building the Biobay GIS," *Geospatial Solutions*, Vol. 15, No. 1, January, 2005, p. 30.

Multisensor Data Fusion for Detection of Unexploded Ordnance, Strategic Environmental Research and Development Program, Cleanup CU-1052, Arlington, Va, n.d.

National Biological Information Infrastructure, Invasive Species Information Node, n.d. As of August 7, 2007:
http://invasivespecies.nbii.gov/

National Park Service and U.S. Air Force, "United States Air Force and National Park Service Western Pacific Regional Sourcebook," Washington, D.C., October 2002.

New York State Adirondack Park Agency data set, n.d. As of June 11, 2007:
http://www.nysgis.state.ny.us/gisdata/inventories/member.cfm?organizationID=508

Office of Management and Budget, *Coordination of Geographic Information and Related Spatial Data Activities,* Circular A-16 revised, Washington, D.C.: The White House, August 19, 2002.

Osman, Trent, "Bobcats & Eagles & Bats, oh my!" *Currents,* Winter 2005, pp. 6–13.

Pluijmers, Yvette, *The Economic Impacts of Open Access Policies for Public Sector Spatial Information,* n.d. As of April 30, 2007:
http://www.fig.net/pub/fig_2002/Ts3-6/TS3_6_pluijmers.pdf

Proffitt, Joe, "Cultural and Biological Survey of Caves at Fort Leonard Wood, Missouri, Assessing the Suitability of Caves for Military Training," *Federal Facilities Environmental Journal,* Winter 2005, pp. 43–53.

Real Property Inventory Requirements, ODUSD(I&E), Arlington, Va., January 2005.

Return on Investment Policies, Concepts, and Methods for Installation Life-Cycle Management, Naval Air Station Patuxent River, Md., May 2000.

RPI Core Data Element Detail Report, Arlington, Va.: ODUSD(I&E), January 19, 2005.

Scott, Russell L., David C. Goodrich, and Lainie R. Levick, "A GIS-based Management Tool to Quantify Riparian Vegetation Groundwater Use," Tucson, Ariz.: Southwest Watershed Research Center, n.d.

"SERDP Launches Research Program at Camp Lejeune, Supports Sustained Military Training in Coastal and Estuarine Environments," *SERCP Information Bulletin,* No. 22, Fall 2004, pp. 6–7.

Silva, E., *Cost-Benefit Analysis for GIS,* NY State GIS consortium, New York, 1998.

Smith, D. A., and R. F. Tomlinson, "Assessing Costs and Benefits of Geographical Information Systems: Methodological and Implementation Issues," *International Journal of Geographical Information Systems,* Vol. 6, No. 3, 1992, pp. 247–256.

Sommers, Rebecca, "FRAMEWORK: Introduction and Guide," Washington, D.C.: Federal Geographic Data Committee, 1997.

Stumpf, Annette, and Suzanne Loechl, *Mitigating Impacts of Encroachment: Tools for Installations and Their Neighbors*, U.S. Army Corps of Engineers, January 7, 2004.

Sun, Zhanli, Brian Deal, and Varkki George Pallathucheril, "The Land-Use Evolution and Impact Assessment Model: A Comprehensive Urban Planning Support System," URISA, February 8, 2005. As of July 24, 2007: http://www.urisa.org/publications/journal/articles/the_land_use_evolution

Sydelko, Pamela J., Kimberly A. Majerus, Jayne E. Dolph, and Thomas N. Taxon, *A Dynamic Object-Oriented Architecture Approach to Ecosystem Modeling and Simulation,*" Strategic Environmental Research and Development Program, Arlington, Va., n.d.

Tang, Winnie S. M., and Jan Selwood, *Spatial Portals: Gateways to Geospatial Information Services,* Redlands, Calif.: ESRI Press, 2005.

The National Map. As of July 8, 2007: http://nationalmap.gov/

"TISP Embraces Geospatial Aspect of Infrastructure Security," *Geo World*, Vol. 5, No. 15, May 2002, p. 18.

URS Group, Inc., *Marine Corps Customized GIS Tools: Inventory and Recommendations*, Morrisville, N.C., October 2004.

U.S. Air Force, *Expeditionary Site Mapping Conops: Strategic,* Washington, D.C.: Pentagon, April 1, 2004.

USAREUR ITAM: Training Support Products and Services Catalog 2005, USAREUR, ITAM, 2005.

U.S. Army Corps of Engineers, *Geospatial Data and Systems*, Washington, D.C.: Department of the Army, EM 1110-1-2909, September 30, 2005.

U.S. Army Environmental Center, "Army Compatible Use Buffer Program: End of Year Report FY05," 2005a.

———, *Fort Bliss Applies Future Range Mission Analysis Process*, Winter 2005b.

U.S. Department of Agriculture, "National Invasive Species Information Center," n.d. As of June 11, 2007: http://www.invasivespeciesinfo.gov/

U.S. General Accounting Office, *Information Technology Investment Management: A Framework for Assessing and Improving Process Maturity*, GAO 04 394G, 2004.

"U.S. Army Training With GIS," *GeoIntelligence*, November/December 2004, p. 45.

U.S. House of Representatives, *Statement of Mr. Philip W. Grone, Deputy Under Secretary of Defense, Installations and Environment*, Washington, D.C., March 2, 2005.

————, *Statement to the House Appropriations Subcommittee on Military Quality of Life and Veterans Affairs*, Washington, D.C., April 7, 2005.

Using Remote Sensing to Detect and Monitor Change on Military Lands, report brief, Strategic Environmental Research and Development Program, Arlington, Va., Conservation CS-1098, 2001.

U.S. Marine Corps, *Marine Corps Installations Strategic Plan,* Washington, D.C., June 28, 2004.

Vance, Tiffany, Jason Fabritz, and Denis Shields, "Mapping Ship Locations and Sensor Data in Near Real Time," *ArcUser*, April–June 2003, pp. 68–69.

Vernon, Adam, "RSIMS: Navy's Eye in the Sky," U.S. Navy Region Northwest, January 14, 2005. As of April 30, 2007:
http://web.soundpublishing.com/index.php/navigator/region/rsims_navys_eye_in_the_sky/

Warren, S. D., H. Mitasova, M. R. Jourdan, W. M. Brown, D. E. Johnson, D. M. Johnston, P. Y. Julien, L. Mitas, D. K. Molnar, and C. C. Watson, *Digital Terrain Modeling and Distribution Soil Erosion Simulation/Measurement for Minimizing Environmental Impacts of Military Training*, Strategic Environmental Research and Development Program, Arlington, Va., February 2004.

Watts, J., W. R. Whitworth, A. Hill, G. I. Wakefield, T. Davo, and L. J. O'Neil, "Vegetation Map Accuracy Assessment: Fort Benning, Georgia," U.S. Army Corps of Engineers, Construction Engineering Research Laboratory, CERL Technical Report 99/76, August 1999.

Weaver, S. D., "Integrated Environmental Site Characterization and Analysis: EQuIS in the Military," EarthSoft, Inc., n.d. As of April 30, 2007:
http://www.earthsoft.com/en/about/articles//ESRI2001_p776.pdf

Westbrook, Cory, and Kinra Ramos, *Under Seige: Invasive Species on Military Bases,* National Wildlife Federation, October 2005.

Westervelt, James D., *Approaches for Evaluating the Impact of Urban Encroachment on Installation Training/Testing*, Champaign, Il.: U.S. Army Corps of Engineers, ERDC/CERL, March 2004.

Willets, Douglas, and Paul Dubois, *Army ITAM GIS: Automating Standard Army Training Map Production,* Fort Collins, Co.: ITAM Western Regional Support Center, Colorado State University, 2003.

Wright, Newell, Vista Stewart, Tegan Swain, and Lynne Shreve, "Using GIS and the Web on Eglin AFB," *Cultural Resources Management*, Vol. 2001, No. 3, March 2001, pp. 9–11.

W. K. Kellogg Foundation, *Logic Model Development Guide*, January 2004. As of April 17, 2007:
http://www.wkkf.org/Pubs/Tools/Evaluation/Pub3669.pdf

Zenilman, J. M., G. Glass, T. Shields, P. R. Jenkins, J.C. Gaydos, and K. T. McKee, "Geographic Epidemiology of Gonorrhoea and Chlamydia on a Large Military Installation: Application of a GIS System," *Sexually Transmitted Infections*, Vol. 78, No. 1, 2002, pp. 40–44.